Urban and Architectural Heritage Conservation within Sustainability

Edited by Kabila Hmood

Published in London, United Kingdom

IntechOpen

Supporting open minds since 2005

Urban and Architectural Heritage Conservation within Sustainability
http://dx.doi.org/10.5772/intechopen.78244
Edited by Kabila Hmood

Contributors
Oluwafemi Akande, Maha Salman, Umali Saidi, Gulsen Disli, Marcus Balah Ryal-Net, Bogda Prucnal-Ogunsote, Erekpitan Omoikhefe Ola-Adisa, Dr. Kabila Faris Hmood

First published in London, United Kingdom, 2019 by IntechOpen
IntechOpen is the global imprint of INTECHOPEN LIMITED, registered in England and Wales, registration number: 11086078, The Shard, 25th floor, 32 London Bridge Street
London, SE19SG - United Kingdom
Printed in Croatia

British Library Cataloguing-in-Publication Data
A catalogue record for this book is available from the British Library

Additional hard and PDF copies can be obtained from orders@intechopen.com

Urban and Architectural Heritage Conservation within Sustainability
Edited by Kabila Hmood
p. cm.
Print ISBN 978-1-83880-881-5
Online ISBN 978-1-83880-882-2
eBook (PDF) ISBN 978-1-83880-883-9

For EU product safety concerns:
IN TECH d.o.o., Prolaz Marije Krucifikse Kozulić 3, 51000 Rijeka, Croatia, info@intechopen.com or visit our website at intechopen.com.

Meet the editor

Prof. Dr. Kabila Faris Hmood was born on April 4, 1962. She lives and works in Amman, Jordan where she is Dean of Faculty of Architecture and Design at Al-Zaytoonah University. She has been a professor since 2004. Dr. Hmood received her PhD from the University of Baghdad, Iraq, in 1996. She completed her doctoral research on the history and theories of Islamic architecture " Proportion and proportional systems of Abbasid Architecture ". She has more than 30 years of experience and has carried out a great deal of research. Dr. Hmood has published more than 35 papers, articles, and studies as well as four books. She is also the recipient of many national and international awards.

Contents

Preface XI

Section 1
Urban Heritage Conservation 1

Chapter 1 3
Introductory Chapter: Heritage Conservation - Rehabilitation of Architectural and Urban Heritage
by Kabila Faris Hmood

Section 2
Reuse of Heritage Buildings within Sustainability 15

Chapter 2 17
Improving Environmental Sustainability in Reuse of Some of England's Churches: Challenges and Options for Sustainable Practices
by Oluwafemi K. Akande

Chapter 3 33
New Additions to Existing Built Heritage and Their Contributions to Sustainable Development: Cases from Ankara, Turkey
by Gulsen Disli

Section 3
Vernacular Architectural Identity within Sustainability 55

Chapter 4 57
Sustainability and Vernacular Architecture: Rethinking What Identity Is
by Maha Salman

Chapter 5 73
Repackaging "Traditional" Architecture of the African Village in Zimbabwe
by Umali Saidi

Section 4
Preservation and Restoration of Historic Buildings 89

Chapter 6 91
Assessment of Selected English Colonial Heritage Features for Sustainable Conservation in Kaduna Metropolis, Nigeria
by Marcus Balah Ryal-Net, Bogda Prucnal-Ogunsote and Erekpitan Omoikhefe Ola-Adisa

Preface

Recently, there has been growing research interest in identity in the architectural and urban environment. Concurrently, there is also a growing interest in sustainability.

This research is interested in achieving an integrated environment to meet human needs as well as guide energy consumption, which preserves both the identity and privacy of architecture. In terms of contemporary architecture, there is an important need to formulate guidelines for planning and designing decisions with regard to the potential of sustainability that is evident in the traditional architecture environment.

There is a contemporary return to the traditional environment, wherein successful and efficient climatic solutions contribute to the rationalization of energy consumption.

As a result, our contemporary architectural styles must be affected by our fundamental traditional concepts. This helps to preserve architectural identity and heritage when our comprehensive architectural works respond to the cultural and natural environment. Contemporary cities should meet the needs of the modern human while maintaining their architectural identity.

In contemporary cities, there is an increasing population growth that causes them to expand. Modern urban and commercial processes have expanded rapidly towards their cities' historical centers, and with the loss of the majority of the cities' master plans, the relationship between the old and the new, as well as its social, economic, and architectural effects, has becomes an issue tackled in many researches. This makes us responsible for developing today's cities, especially our historical cities, for the future. As a result, the cities of the future must be based on a foundation that ties together the past, present, and future through an advanced connective chain that meets all humans' requirements in any era. Cities with no memory miss their past and, as a result, can't dream of building their future, which makes them vulnerable and prone to destruction.

There is importance in finding a clear visualization for planning the architectural design for historical cities in the present and in the future. This plan puts us on the path of sustainable development, which satisfies the needs of the present and protects the future. This is why it is important to study "urban conservation."

Chapters of this book attempt to answer a number of questions, such as what is urban conservation? How do cities of historical depth deal with modern planning policies? What are our policies for conservation, rehabilitation, restoration, and preservation of the urban fabric of historical cities and heritage buildings? How can we explain urban conservation, sustainable heritage, and historical sustainability?

The preservation of cultural and architectural heritage requires the participation of specialists from various disciplines, including architects, urban planners, economists, sociologists, archaeologists, artisans, interior design experts, law experts and building legislators as well as specialists from engineering disciplines related to structure and infrastructure.

This book consist of four sections. Section 1, "Urban Heritage Conservation," includes the introductory chapter, "Heritage Conservation; Rehabilitation of Architectural and Urban Heritage." I try in this chapter to explain some concepts and procedures for urban conservation. It is an attempt to come up with guidelines for the planning and design of contemporary architecture with respect to sustainability. This sustainability potential that emerges in the traditional architectural environment helps preserves its past and maintains its identity.

Section 2, "Reuse of Heritage Buildings within Sustainability," includes two distinguished chapters. Chapter 2, "Improving Environmental Sustainability in Reuse of Some of England's Churches: Challenges and Options for Sustainable Practices," aims to investigate current retrofitting strategies and interventions in heritage buildings. The purpose is to assess current performance through the viewpoint of energy efficiency. The chapter adopts a pragmatic analytic and comparative approach and methodology to investigating retrofitting interventions in the reuse of England's listed churches. Chapter 3, "New Additions to Existing Built Heritage and Their Contributions to Sustainable Development: Cases from Ankara, Turkey," uses data from a new survey of historic buildings in Ankara with new exterior or interior additions of previous restorations in order to reveal the quality, compatibility, and/or incompatibility of contemporary new additions. It also explores contributions to sustainability using five case studies of historic buildings in the Ankara, Ulus district.

Section 3, "Vernacular Architectural Identity within Sustainability," also includes two chapters. Chapter 4, "Sustainability and Vernacular Architecture: Rethinking What Identity Is," aims to redefine what identity is as a concept. It also discusses the impact of globalization on contemporary architecture, especially on regions with rich heritage and unique culture such as the Arab world. The chapter examines the emergence of "local identity" as a reaction to the globalization of cultural values, uniform architectural styles, and stereotype patterns through discussing sustainability as a motivation for identity in culture and architecture.

Chapter 5, "Repackaging 'Traditional' Architecture of the African Village in Zimbabwe," architecturally and onomastically explores selected outdoor spaces in Zimbabwe. Repackaging of these spaces is being done through construction of buildings with connections to traditional architecture as well as respective traditional naming of such, symbolizing a return to African values. The names of the spaces communicate African sensibilities about the African aura of the traditional village and at the same time sustainably conserve the architecture authored by African heritage.

Section 4, "Preservation and Restoration of Historic Buildings" includes Chapter 6, "Assessment of Selected English Colonial Heritages Features for Sustainable Conservation in Kaduna Metropolis- Nigeria." This chapter sets out to establish heritage significant values, determine the relative influential index of deterioration, and assess the suitable sustainable heritage conservation options within the colonial Nigerian town of Kaduna.

I would like to thank all the authors for their valuable contributions to this book. I would also like to thank IntechOpen's Author Service Manager, Ms. Maja Bozicevic, for her cooperation and competence in the management process and her help in bringing this book to publication.

Kabila Faris Hmood
Dean of the Faculty of Architecture and Design,
Department of Architecture,
Faculty of Architecture and Design,
Al-Zaytoonah University of Jordan,
Jordan

Section 1

Urban Heritage Conservation

Chapter 1

Introductory Chapter: Heritage Conservation - Rehabilitation of Architectural and Urban Heritage

Kabila Faris Hmood

1. Introduction

> *Without the understanding and support of the public at large, without the respect and daily care of the local communities, which are the true custodians of world heritage, no amount of funds or army of experts will suffice in protecting the cites.*
> *—Mr. Koïchiro Matsuura, Former Director-General of UNESCO*

For the time being, we are interested in preserving the heritage because of the loss of culture, heritage, and inherited buildings. For more than 100 years ago, there has been an increasing research interest in the preservation of architectural heritage, restoration of monuments and historic buildings, urban renewal policies, traditional architecture and its identity, etc. On the other hand, there is a growing interest in sustainability.

On this basis, there is an important need to formulate guidelines for planning and designing decisions of contemporary architecture that belongs to the sustainability and appear in the traditional architectural environment, which preserves its past and maintains its identity. Modern cities must meet humans' needs and preserve their urban identity. Planning policies must be developed in order to achieve the sustainable development goals, which meet the needs of the present and protect the future rights that preserve the past and constitute a strong basis for the present and future.

2. What is the purpose of this book?

This book tries to answer many questions, in order to reach to the required conclusions and recommendations. The questions are as follows: What is the urban conservation? How do cities with historic depth deal with modern planning policies? What are our policies to preserve and rehabilitate the urban fabric of historic cities and heritage buildings? How do we explain urban conservation, sustainable heritage, and historical sustainability? How can we add new additions to the current building heritage and what is their contribution to sustainable development? What is the role of the organic additives in heritage structures? Shall we rethink of identity, through the study of traditional architecture within sustainability? Our book meets human needs and energy consumption and maintains the architecture identity with respect to sustainability.

3. What is heritage conservation?

The preservation of identity in the cities has a great relationship with the correct planning policies that relate to heritage buildings and traditional architecture,

 IntechOpen

where the traditional cities contain organic fabric with narrow, twisted, or broken roads with incomplete visual axes. The streets in the traditional cities are shaded and sometimes even roofed, as in Jerusalem, Damascus, Aleppo, etc. In addition to the inner courtyards, this made them an inward-looking buildings [1], in contrary to what we see those days in modern cities.

3.1 What is heritage preservation?

It is important to preserve historic cities and heritage buildings and revive them. The policy of reuse and rehabilitation of these legacies with selecting the right contemporary functions is a successful step on a secure path to save what can be saved from the heritage (**Figure 1**).

The World Heritage Convention recognizes that heritage can be defined as "monuments, group of buildings, and sites." There is a wide range of styles including urban cities, archeological sites, industrial heritage, cultural landscapes, and heritage roads. This increases the scope of places and landscapes that must be managed by heritage managers. By this definition, heritage is not only buildings such as worship places or fortresses and castles, which were considered as independent

Figure 1.
Amman City, Jordan (past and present).

places of its own not related to its urban environment. Today, heritage represents the whole environment that is influenced by its interaction with humanity and is therefore qualified to be recognized as a heritage [2].

There are several definitions of architectural and urban heritage preservation. For example, Elias considered that "conservation is a set of procedure which guarantee the protection of the architectural and historical characteristics of the important old areas and buildings, such as repairing, removing erosion and sedimentation signs, and securing acceptable standards for the protection of open spaces and squares. These emphasize the importance of urban preservation and the need for the functional continuity of these areas and buildings" [3]. This means preserving the heritage while insisting on keeping up and responding to the modern urban fabric of the city as a whole.

Feilden defines the cultural heritage by making the architectural heritage an important part of it, because it is "the physical image and material embodiment of unique human components in which man is the basis of creation, creativity and production" [4]. Since the ruined or heritage building cannot be separated from its urban surroundings, therefore, urban conservation is what we should seek.

Today, in contemporary cities, there is a rapid population growth leading to the expansion of contemporary cities, and with the loss of most master plans of cities, the relationship between ancient and contemporary has become an issue that has been addressed in many researches. This makes us responsible for developing today's cities, especially our historic cities for the future. As a result, future cities must be based on the past, living in the present, and evolving in the future through a coherent chain that meets the needs of all people in any age.

Cities that lack memory, away from their past, and do not preserve their heritage and identity cannot build their own future, which make them vulnerable to destruction.

This leads us to the importance of studying the term "urban conservation."

Urban conservation is "a process that focuses on the quality of urban environment by developing a program to protect cities and urban areas from environmental and optical pollution" [5].

The definition highlights the importance of restoration as an essential part of conservation procedure, regardless of the way or the area covered by preservation, whether it is a single building, urban fabric, or an entire city. They are maintained through different methods depending on the state of the building, the factors that led to the deterioration, its original or required function, and the proposed function.

We must understand and pay attention to preserving the traditional urban fabric, at least in the case of heritage loss and the addition of the modern element, as what happened in Leipzig. It happened also in Vienna, with a clear harmony between the old and the modern by preserving the important monuments and morphology of the city (**Figure 2**).

Consantinah – Algeria | Old City of Baku – Azerbaijan | Ulus – Ankara Turkey | Sidi Bu Said – Tunisia(Tunis)

Figure 2.
Urban preservation by figuring out the harmony between the old and the present.

We found a beautiful harmony between the old and the present, as well as preserving this harmony in Danzig (located Northern Poland) from war damages [6]. In Baku, Azerbaijan, the preservation of the internal order of the urban fabric was observed. This means the preservation of the urban structures, roads, and historic buildings. In Baku, it allowed us to find a new urban pattern, linked successfully to the old type.

Accordingly, urban conservation in historical cities can be achieved during its expansion, in an attempt to meet the requirements of the age. This can be achieved through applying the urban renewal policies that attempt to preserve the urban fabric and the city's inherited structure, as well as the urban flow of modern requirements, without forgetting the harmony of the architectural type of buildings and surroundings.

Heritage is a moral and material entity. Therefore, preserving it is a necessity in urban renewal policies of cities in all countries of the world, especially cities with historical depth, to achieve historical sustainability. The policy of maintaining their differences can stop the deterioration and the decline of the traditional and historical urban fabric [7].

Countries with special cultural heritage provide a humanitarian service to the world community and its society in particular. Therefore, many international organizations are interested in preserving the heritage and providing emotional, scientific, and material support to many countries such as UNESCO, Arab Towns Organizations, Aga Khan Institute, and others.

There are many countries through UNESCO which provide material and scientific support in order to save heritage in other countries of the world [8].

4. Conservation methods and policies

4.1 Methods and policies of conservation

Conservation methods are consolidation, reproduction, reconstruction, preservation, deterioration perversion, rehabilitation, and restoration.

The cultural awareness of the importance of heritage preservation in general, architectural and urban heritage in particular, was more evident after World War II in the twentieth century. Several policies were applied in the old cities and historical areas of Europe after World War II, including the removal and compensation policy. There is a reconstruction policy which deals with the replanning and constructing the areas that have been completely destroyed.

In addition to the conservation and preservation policy, which is concerned with the urban, social, and economic environment in order to preserve the buildings, the fabric and the special urban character of the old city, the use of inherited buildings, and their fabric to suit the age without affecting building exterior represents the policy of reuse and refunction. Once the adopted policy maintains to preserve the building, which extends to the surrounding area by intervening with the inherited buildings (renovation, restoration, maintenance, and reuse) and to its historical surroundings, we have to adopt the rehabilitation method. In addition, we have restoration policy, and when we put the criteria and conditions to ensure the survival of these buildings in the urban framework, this will call the protection policy, which are related to the restoration and renewal policy of inheritances [9].

Keeping up with the architectural and urban heritage is achieved through the process of rehabilitation and reuse within a different age other than the one that produced these inheritances. This can be achieved through selecting a suitable contemporary function, which is not impossible, but it is very difficult and complex, where several factors overlap in order to get the balance to reach an acceptable outcome and achieve a successful experience.

The specialists must balance between these factors and influences and consider them, starting from the nature of the inheritance, its internal spaces, their size, distribution and its absorption of new activities, and the value of historical and architectural heritage. The type of the proposed function comes from the origin of the adopted function of the inheritance that is derived from its location, the time of construction, and the most important events associated with it in addition to its architectural and urban style. The ownership of the inheritance is considered to have a significant impact on the process of its functional rehabilitation, since the responsible authorities cannot preserve the heritage when it is considered as private property. The selected function should be homogeneous with the surrounding urban fabric and should be achieved with the least amount of modifications [10].

4.2 Rehabilitation of heritage buildings

The preservation of heritage, and in particular architectural heritage, goes through several stages, starting with documentation, collection of enough data and information about each heritage, documentation by photos, architectural plans, tables, etc.

By teaching the following course, "Preserving Architectural Heritage," I realized the importance of connecting the theoretical aspect with the practical one, where the student learns the scientific stages that he/she should follow for the success of the heritage preserving process.

After placing students into groups, they were asked to submit a conservation study of a preserved single building or a group of historical or heritage buildings without proper contemporary use or the lack of documentation or those that deserve to be preserved but neglected. Students were then asked to document, study, and analyze them to choose a contemporary function, which allows heritage building or group of buildings to achieve heritage sustainability in the current age. In the case of rehabilitated heritage buildings, the present function is evaluated and we propose another one to be developed according to the standards and the scientific foundations, which were adopted in this regard. Anyway, it is a direct means of documenting these achievements and preserving them from loss, and reviving them for our contemporary life and for the future of generations, which preserves the nation's memory.

The heritage preservation process of inherited models is provided through:

- Registering
- Describing
- Taking photos to preserve its features
- Measuring its size to draw its plan accurately

This is done by conducting a comprehensive field survey of the inherited stocks with interviews of specialists and persons involved in the inherited architectural or urban model. In addition to relying on the literature and sources that can provide us with data, information, and documents that make preservation (and often rehabilitation) an accurate and calculated process, which increases the value of architectural and urban heritage and is not a reason to harm it if it is based on inaccurate data and information, the results of this documentation are proved in tables, records, and maps to facilitate the heritage whenever necessary, whether at the stage of the preserving decision of those inherited models or while maintaining and restoring them to prepare them for new or previous use after making the inheritance suitable and compatible with our age.

4.3 Choosing appropriate function

At this stage, there are a lot of difficulties linked to multiple aspects to ensure the proper selection of the new use; many of the traditional models suffered from damage and loss due to poor preservation and bad selection of function. The traditional model site, its architectural value, its original function, distribution of its spaces, their size, services, and its ownership are the factors behind the selection of the new suitable function.

And preparing the historical model for the its new use, based on the fact that space, as Schulz sees, meets the needs of man in a way that creates a close relationship with his environment, and so space is embodied in architecture, which achieves useful and symbolic purpose. The first useful purpose can be achieved by meeting human's needs and his various activities within a comfortable environment, and the second purpose is achieved through the meanings and expressions that space provides to its occupants.

The task of selecting the appropriate function in the old heritance rehabilitation is not easy. Some difficulties may arise from the fact that the selection process is subject to multiple factors and different influences. The team of specialists that maintain and rehabilitate must balance all of them. The type of the function often is derived from the origin of the adopted function of the inheritance and that came from its origin, time of its construction, and most important events related to it, in addition to its architectural and urban type.

The ownership of the inheritance has a significant impact on its functional rehabilitation, especially when it is a private property. The owner should be aware of the value of the inheritance and preserve it to achieve cooperation with the responsible authorities (either governmental or otherwise).

The selected function should suitably relate to the inherited environment and with minimal changes to harmonize between the new proposed use and with the old use, so that they will pay a reasonable cost for that change. In addition, that process will be done with the least cost, effort, and time. We should take into account the variable usage from time to time for the inheritance.

An example of heritage building rehabilitation is the building of the Consulate of Tuscany (**Figure 3**). It was built in the second half of the eighteenth century under the reign of Ali Pasha al-Qara Manli, who ruled Tripoli, Libya, from 1754 to 1793. It is considered as one of the most beautiful houses in Tripoli, known as Hosh Al Hareem (courtyard of women). It was used in the Second Ottoman Period as the Consulate of Tuscany. This building was restored and then opened in 1994 to be a building for folk collections and historical fashion in the name of "Tripoli Historic Exhibition of Folk Collections and Historical Fashion."

We have also Al Azem Palace in Damascus, Syria (**Figure 4**), which is one of the most impressive residential buildings in Syria, built in the eighteenth century, and was originally the residence of the ruler. Later, the palace became the museum of art and folk traditions in Syria. It has a traditional style and is divided into two parts, which consists of the official area, bathrooms, and exterior buildings. Each room embodies a particular social situation. Wax dolls are used in the museum to embody the characters that tell the story of the place and time.

Figure 3.
The central courtyard of the Consulate of Tuscany, which is now the Historical Exhibition of Tripoli, Libya [11].

Figure 4.
The distinctive architecture in Al Azem Palace, Damascus (which is a residential building reused to be a museum).

4.4 Space modifying with the new function

This stage was formed through the changing and modifying of the internal and urban spaces and their designed elements. In order to employ it in line with the revival of the heritage model, the users' needs and the new function requirements, at the level of all interior design elements and without affecting the value of the historical, architectural, and esthetic heritage model, and the different services that are required by the age when the model was rehabilitated.

Any defect in the space rehabilitation process is a failure in the process of preservation and rehabilitation, which will result in the same damage of not selecting the appropriate function, accurate rehabilitation, and in maintaining, registering, and documenting the traditional models. These spatial adaptations are not only at the level of the internal space elements of the architectural heritage, but also include the required spatial adaptations in the open spaces, in the inherited urban fabric, and the required changes to suit the chosen functions of this inherited urban area. The suitable choosing of new function is based on its centrality within the fabric of the city, its relationship with the surrounding planning units, the historical roots and original functions, as well as the impact of the new functions proposed to rehabilitate the inherited urban area, and the nature of the heritage buildings. In addition, of its age and the state of its construction and other influences that is the formation of space, then adapting it in line with the requirements of the age and the users' needs of the inherited urban fabric.

After restoration, maintenance and space modifications   During restoration, maintenance and space modifications

Figure 5.
An old building that has been preserved and rehabilitated as a hotel (Libya).

Figure 6.
Traditional Baghdadi houses that have been rehabilitated for administrative use.

In Tripoli, Libya, we can find the Izmeet Hotel, founded in 1816 and used since its establishment in the First Ottoman era as a hotel and a forum for caravan traders between Europe, North Africa, Countries of Africa, and beyond the Sahara. Izmeet Hotel is located in Bab El Bahr, near the port of Tripoli, not far from the Roman Marcus Aurelius, the Roman ruins of the city.

This hotel has been renovated, maintained, designed, furnished, reused, and rehabilitated. It operated for 5 years. It consists of two floors with 10 rooms and 14 suites provided with all services. It includes a dining hall, an Arabic session, a coffee shop, and a room equipped with Internet. The whole parts of the hotel are furnished with traditionally handmade pieces (**Figure 5**).

Modifications can be made on some traditional houses to be used for administrative purposes, as is the case with two houses in Baghdad, Iraq (**Figure 6**).

One of the heritage buildings that has been used and rehabilitated with a new function is the Khan Murjan building from the Mughal period. The Khan is famous for the beauty of its architecture, design, stone decoration, and the uniqueness of its arches. It is one of Baghdad's most famous Khans. It was rebuilt by the Department of Archeology and Heritage and reused as the House of Antiquities and Heritage (administrative use) and then as the Museum of Islamic Art.

Khan Murjan is now a wonderful tourist restaurant consisting of two floors; the first floor contains 22 rooms and the second floor contains 23 rooms. The ceiling consists of eight huge arches. It includes a series of walls and vaults with different shapes and dimensions interspersed with windows. This type helps to provide natural lighting and reduces the amount of pressure caused by the weight of the arches [12] (**Figures 7** and **8**).

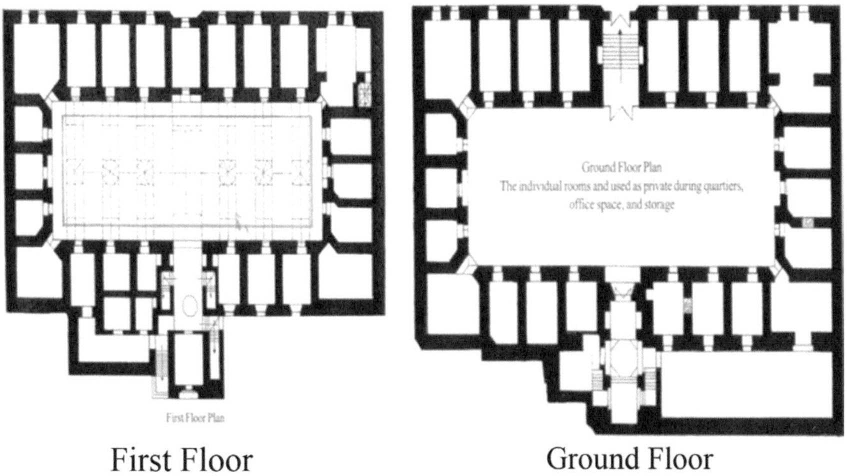

Figure 7.
Khan Murjan in Baghdad, Iraq.

Figure 8.
Khan Murjan was built since Mughal period in Baghdad, Iraq.

As for the most appropriate contemporary functions that are selected to rehabilitate the heritage for preservation, the most specialized specialists in architecture, history, heritage, archeology, arts, maintenance, and interior design, who have distinct interests in preserving the inheritance have agreed that:

- Cultural functions (museums, libraries, art galleries, exhibition halls, scientific and cultural centers, etc.) are at the top of the list of selected functions because they extend the age of the inheritance and do not harm it in the future. They are also compatible with the historical roots and the symbolic and architectural values of heritage.

- These functions are often cultural tourism followed by tourist functions:

  - Educational functions

  - Residential functions

  - Entertainment functions

  - Administrative functions

- Commercial functions
- Health functions

Industrial function is one of the most harmful functions for the inherited buildings and inherited urban fabric, because this function results in damage and cracks in the body of the inheritance that causes loss over time. In the Aleppo preservation experience, the cultural tourism uses achieved the highest use, and there is almost no industrial function in the old fabric.

5. Conclusion

Attempts to preserve historical cities and cultural heritage are very few compared to the historical richness of the world. We highlight the historical identity when preserving our heritage. It is necessary to increase the degree of interest that strengthens the balance between historical and heritage wealth and the achievements of preserving it.

The preservation of heritage in all its forms poses a number of unique problems. The issue must consider in terms of the cultural, economic, and political values of the society in which it exists. The problem of financing is one of the most important obstacles to accomplish this task. It is necessary to cooperate with the governmental and private sectors at the local level. It is also possible to involve the international institutes interested in this aspect, because heritage is a human achievement.

The participation of people in preserving their heritage, identity, and awareness of its importance is considered as an effective factor in achieving the sustainability of these inheritances through a successful program of sustainable development. The legislation and administrative procedures are important for the success of conservation operations.

This can be achieved through granting people incentives and supporting them to participate in maintaining to encourage them stay in the city and preserve it. It is necessary to study the environmental importance of the old city, open spaces, gardens, and infrastructure, and involve the population in developing protection plans in cooperation with the competent authority. It is also useful to encourage tourism to the old city without harming it. This makes the population feel that they are materially benefiting from the tourists and that their cultural property deserves maintenance and protection because it is part of the world heritage.

Therefore, this awareness should start from people, society, and enterprise to achieve our goal of preserving heritage buildings and historic cities, in order to preserve the cultural and urban identity. We should also bear in mind the importance of the contribution of heritage protection in sustainability and sustainable development, which varies from case to another, during the three "pillars": the social dimension, the economic, and the environmental dimension.

Finally, we must emphasize that the overlap of several aspects is needed to achieve a successful preservation experience of a city. These aspects are planning, designing, housing and construction, laws and legislation as well as the field of tourism, preservation of the architectural heritage, infrastructure, services, environment and community participation with media to activate the sense of responsibility, awareness of the importance of the city, and its cultural and historical value.

Conservation should also include traditional and local architecture, as it always places the option of preserving through rehabilitation and letting the inherited buildings up-to-date.

Author details

Kabila Faris Hmood
Department of Architecture, Faculty of Architecture and Design, Al-Zaytoonah University of Jordan, Jordan

*Address all correspondence to: khmood85@gmail.com

References

[1] Ahmad NHH. The homogeneity of urban planning with the historic context in the Arab City. Architectural Scientific Journal. 1988;**4**:80

[2] Managing Cultural World Heritage. World Heritage Resource Manual. Vol. 7. Paris, France: United Nations Educational, Scientific and Cultural Organization; 2013. p. 12

[3] Elias IJ. Foundations of urban renewal of heritage textile [Unpublished Master thesis]. Baghdad: Department of Architectural Architecture, University of Technology, Baghdad; 1989. p. 4, 30

[4] Feilden B. Introduction into Conservation. Rome: UNESCO Press; 1979. p. 34

[5] Hmooud FK. Arab Architectural and Architectural Innovations. Amman, Jordan: Dar Al-Warraq; 2011. p. 30

[6] Cohen N. Urban Planning Conservation and Preservation. New York: McGraw-Hill; 2011. pp. 24-94

[7] Hmood FK. Architectural and Architectural Heritage in the Arab World. Amman, Jordan: Dar Al-Warraq; 2004. p. 114

[8] Ibraheem MA-A. Preservation of heritage, national and cultural requirement. Architectural Scientific Journal. 1988;**4**:25

[9] Al-Qasim A, Al-Taher R. Towards sustainable development of old cities. In: Al-Qaziri SK, editor. The Preservation of Old Cities. Benghazi: Organization and Administration of the Old City of Benghazi; 2007. p. 215

[10] Hmooud FK. Arab Architectural and Architectural Innovations. Amman, Jordan: Dar Al-Warraq; 2011. pp. 76-77

[11] Jubran MM et al. Guide to the Old City of Tripoli. Tripoli: The Project of Organizing and Managing the Old City; 2002

[12] Joseph S. The History of the Art of Iraqi Architecture in Different Areas. Baghdad: The Ministry of Culture and Information; 1982. pp. 522-527

Section 2

Reuse of Heritage Buildings within Sustainability

Chapter 2

Improving Environmental Sustainability in Reuse of Some of England's Churches: Challenges and Options for Sustainable Practices

Oluwafemi K. Akande

Abstract

Considering the spontaneous growth in retrofitting practices, existing buildings, particularly those of historical significance are being transformed using a wide range of interventions. However, the pervasiveness of these interventions constitutes a serious challenge to retrofitting heritage buildings. The aim of this paper is to investigate current retrofitting strategies and interventions in heritage buildings. The purpose is to assess current performance through the viewpoint of energy efficiency. The paper adopted pragmatic analytic and comparative approach and methodology to investigating retrofitting interventions in the reuse of England listed churches. A top down approach method of data collection was employed to collect energy use data from monthly utility bills and meter printer outs from selected buildings. Findings show that in terms of energy performance, the majority of the surveyed buildings are currently under-performing. Recommendations for low energy use interventions for operational management of retrofitting projects were proposed. It concluded that the low operational energy use should be a key priority for effectiveness in any proposed retrofitting intervention on heritage building projects.

Keywords: heritage buildings, low energy, energy performance, rehabilitation, sustainability

1. Introduction

In Europe, 40% of the total energy use and 36% of CO_2 emissions originate from the building sector. The aim of the European Union (EU) is to reduce its greenhouse gas (GHG) emissions by 80% by 2050. This is reflected in all the EU Directive ([1–3], pp. 1–56). Thus, the building stock plays a major role in achieving the 20-20-20 strategic targets. However, unless other avenues are explored to reduce the environmental footprint attributed to the existing building stock, the EU target may not be met. Due to climate change protection, energy consumption is required to be checked through greater efforts and concentration on existing buildings. According to ([4], pp. 1–87) building professionals need to provide more energy-efficient refurbishment of existing buildings to bring them to modern sustainability standard. However, the possibility lies in adapting and retrofitting existing buildings to the optimum energy efficiency standard ([5], pp. 1–62). The concept of sustainable development could be applied to sustainability of Heritage buildings as any interventions to extend its lifespan without compromising its future and context. In conservation literature, different interventions are found and the term is used as a collective noun which encompasses any works to change, modify, repair or maintain the historic environment in good condition as well as preserve its historical and cultural value or significance. This is discussed in ([6], para 6.11) as "Interventions and Judgement" and as the "action that has a physical or spatial impact on a historic building or its setting."

Generally, the terms used to describe interventions create overlap with other definitions and are often used as synonyms. Prudon [7] defined retrofitting as the act or process of making possible a compatible use for a property through repair, alterations, and additions while preserving the features which convey its historical, cultural and architectural values. More often, retrofitting could involve modernization and change in use (i.e. adaptive reuse). It is considered by [8] as the best way of preserving buildings. Meanwhile, ([9], pp. 155–157; [10], pp. 143–156) referred to retrofitting as similar to the term refurbishment, which does not only extend the life of a building, but also define a new purpose for it according to the demands of modern life. Thus, in this paper, retrofitting is used interchangeably with adaptive reuse.

2. Adaptive reuse of built architectural, cultural heritage

Several scholars ([11], pp. 287–94; [12], pp. 543–56; [13], pp. 529–42) have acknowledged the growing trend in the move to building re-use and adaptation in the built environment. They suggested that some form of adaptation might be able to reduce the impacts of climate change on the built environment. Other authors ([14], pp. 88–103; [15], pp. 43–66) have posited that adaptation is an effective strategy for improving the sustainability of existing buildings along with its potential of giving extension of life to a building. The authors argued that by reusing existing buildings, lower energy consumption can be achieved thus making a considerable contribution to sustainability. With the advantage and possibilities of extension of life for buildings, adaptive reuse could also play a significant role in meeting the growing demand for regeneration of the built environment ([16], pp. 554–70).

According to Langston et al. ([17], pp. 1709–18) adaptive reuse has become an essential strategy to improve the environmental, financial and social performance of buildings. The environmental concern in adaptive reuse of buildings has been acknowledged by other researchers [18–20] in historic preservation. Therefore, it is seen as vital to sustainable development ([21], pp. 1709–1718) and considered applicable to the present climate change adaptation agenda. It is acknowledged that reusing existing facilities are related to sustainable development and in order to promote sustainability within the built environment, many buildings of cultural and historical significance are being rehabilitated. However, little attention is given to improving their operational energy performance. Several factors have since been advanced to be driving the adaptive reuse of buildings such as its value as a practical approach for delivering buildings for new uses, cost-effectiveness and rising energy costs. Latham ([18], p. 8) noted that adaptive reuse is cheaper than new development as it is a way of banking the built environment. Further, he argues that "transforming uneconomic buildings using green materials have the potential to enhance efficiency, comfort and life span of the building". Meanwhile, van't Hof cited in

([15], 43–66) opined that economic considerations have been the major driver behind adaptive reuse although other motives might have also been considered.

2.1 The drivers of adaptive reuse of listed churches

In Europe, many religious heritages are under threat and the buildings are often ill adapted to the needs of modern society [22]. This is because a lot of these buildings are reaching the end of their useful life and in most cases do not respond well to contemporary needs. As a result, they are often less desirable to occupy, can remain empty and ultimately deteriorate. In England, three quarters of 16,000 parish churches of England are listed as buildings of architectural and historic interest. The churches listed Grade I comprises about 45% of all England's buildings (i.e. Castles, mansions, banks, railway stations, etc.). However, with the declining congregational sizes, a number of these buildings are becoming less used and closed for worship (**Table 1**). Thus, one of the drivers for retrofitting of heritage buildings is redundancy.

It can be seen from **Table 1** that from 1969 to 2014, over a thousand church buildings have been closed and considered for alternative use. Perhaps, this is as a result of historical factors such as population shifts, changes in religious practice and habits, or even the construction of new buildings for religious use. According to [24], redundant buildings are buildings that have reached the end of their original working lives, but often have huge potential to be adapted to economically viable new uses. It should be recalled that one of the Council of Europe's statutory duty is to safeguard the ideals and principles of the common heritage of member states and to which religious buildings bear witness ([25], p. 1). This brings to fore religious buildings' importance in terms of their architectural and historical

Alternative use	1969–2010	2010–2014
Adjuncts to adjoining estates	7	0
Arts, crafts	20	3
Civic, cultural or community	150	16
Educational	35	3
Light industrial	11	1
Monument	147	13
Museums	16	0
Music or drama	15	0
Office or shopping complex	58	4
Parochial or ecclesiastical	75	7
Private and school chapel	22	2
Residential	276	36
Sports	15	1
Storage	21	2
Worship (Christian bodies)	160	23
Other	5	0
Alternative use sub-total	**1033**	**111**

Source: Church Commissioner Report [23].

Table 1.
The future of closed church buildings since (1969–2014).

significance and their longstanding concern for the integrated conservation to ensure a future for the past. In the UK, heritage buildings such as churches and farm buildings are being reused. Meanwhile, due to their population (**Figure 1**) more listed churches are converted to alternative use and/or demolished.

When a religious building is no longer viable; efforts should be made to ensure a future use, whether religious or cultural, as far as possible. However, this should be compatible with the original intention of its construction and as such should be carried out with the understanding that a church or any other major religious building is often the focal point and central feature of a community and a local landmark. While there are numerous successful examples throughout Europe of the preservation and protection of redundant religious buildings, through their sensitive adaptation to new uses, thus this paper poses a challenge to responsible stakeholders (e.g. churches, government, local authorities and heritage building professionals) and other heritage building experts to (i) consider effective measures to preserve redundant religious buildings and secure wherever possible their appropriate future use; (ii) promote projects for reuse and re-adaptation not incompatible with the original function of the building and do not cause irreversible alteration to the original fabric; (iii) encourage a more imaginative use of existing religious buildings as well as (iv) encourage the research necessary for the continuous upkeep of religious buildings. This challenge provided the motivation behind this research and action required to implement strategies aimed at promoting energy efficiency and limiting energy consumption as a fundamental aspect of reuse of built heritage. Thus, informed by the aforementioned challenge, this paper seeks to investigate current retrofitting strategies and interventions in the reuse of heritage buildings with a view to assess the current performance of LCBs through the viewpoint of energy efficiency.

The adaptive reuse of church buildings becomes significant in conservation fostered by the economic benefits associated with tourism they could generate

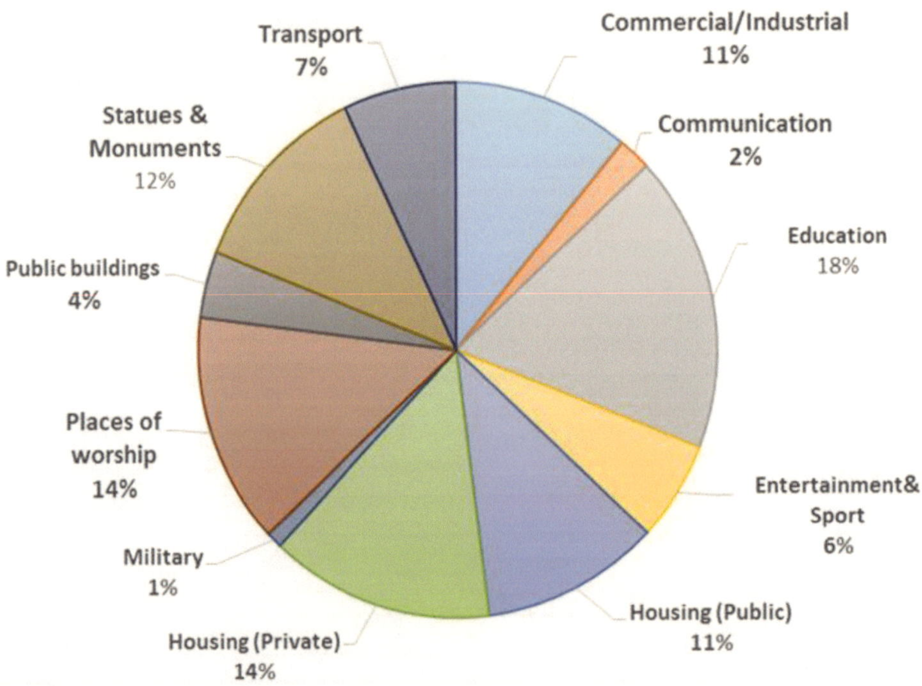

Figure 1.
Post war listed building types in England. Source: Author's survey [30].

[26, 27]. However, retrofitting of these buildings is faced with the challenges of meeting the global challenge of coping with climate change. A major challenge for heritage buildings is how they can successfully be rehabilitated and made energy efficient at a time that the need for their renovation and re-use appears to be urgent and make them fit for the twenty-first century. "In Europe, the legislation concerning this area is based on the certification of energy efficiency, developed in the early 1990s as a primary strategy to reduce energy use and carbon emissions as well as the energy policy adopted in 2007, called Horizon 20-20-20" ([28], pp. 1294; [29], pp. 1493–1502). However, low energy use as a key contemporary demand for better standards of living and as a response to climate change has not yet been extended to the retrofitting of LCBs. The scale of the problem is exacerbated by the fact that churches are difficult to modify to meet up with current energy efficiency standard. Nonetheless, users of rehabilitated church buildings also need to have healthy and thermally comfortable internal environments at an affordable installation and running cost. The problem is that certain restrictions deriving from the specific historic character do not permit major interventions to improve the building's energy performance. Indeed, when dealing with protected buildings of significant architectural merit, altering the building envelope will be prohibited. This difficulty is partly due to the nature of the materials from which they were built as traditional buildings; which affects their thermal performance in terms of heat loss requiring significant updating.

LCBs continue to present a retrofit dilemma all of their own. For instance, the fabric of heritage buildings functions in a certain way due to the way they were built. Thus, makes it challenging to improve their fabric thermal performance. However, to avoid degradation of their fabric they should be preserved because of their breathable elements. Thus, a clearer understanding of their values and needs must be found so that an appropriate intervention can be adopted. Specifically, the challenges of rehabilitating Heritage buildings could be attributed to several factors such as heritage factors, embodied energy, economic factors and building factors (**Figure 2**).

In addition, it is difficult to understand their current energy performance. LCBs traditionally have solid walls, meaning they are 'hard to treat buildings' and hard to

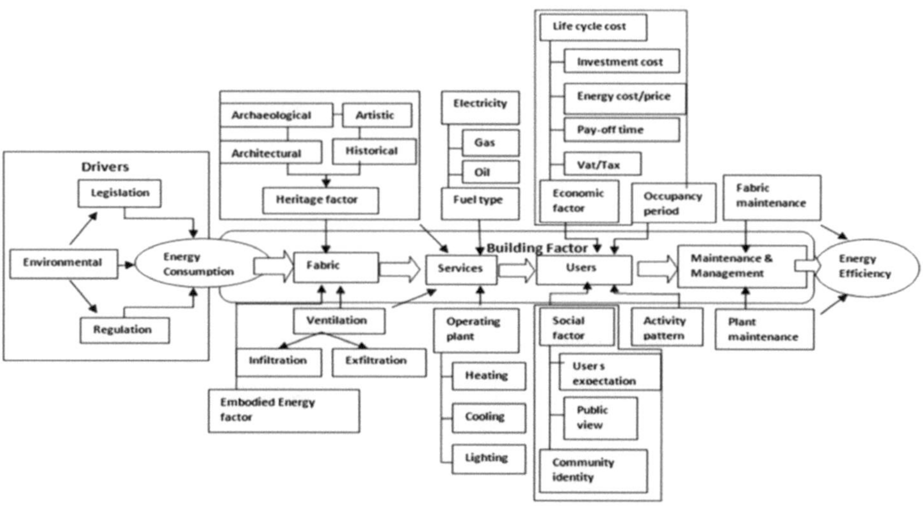

Figure 2.
The drivers and the challenges of retrofitting and environmental sustainability of heritage buildings. Source: Akande [30].

deal with when alterations are locally permitted. If the fabric of the building cannot be readily improved, the question that this study seeks to answer is what options are available to reduce the energy consumption of Heritage buildings with specific reference to investigate operational energy performance involving existing adaptive reuse of LCB projects. Thus, this paper's aim is to investigate current retrofitting strategies and interventions in heritage buildings. The objective is to assess current performance of the existing reuse of heritage buildings through the viewpoint of energy efficiency.

3. Research development and methodology

In developing the research methodology an interview survey was devised, comprising a semi-structured interview. According to [31] there are two types of interviews relevant for subjective surveys. One of them is exploratory or in-depth interviews normally used for surveys. Retrofitting of LCBs was targeted for the survey and adopted for this study. The focus of this study was developed from part of a doctoral study on energy management in the reuse of LCBs. A set of questions was formulated in accordance with the objective of the study and a draft questionnaire was prepared and piloted. Comments were received back from the piloted questionnaire after which a number of redrafts of the questionnaire were undertaken. The final version of the questionnaire contained 85 questions which covered an extensive range of operational issues in retrofitting projects. The questionnaire was designed to address different areas of investigation categorised into six different parts namely: building characteristics, energy using equipment/systems, how the equipment is used, energy used, energy management strategies and user behaviour.

3.1 Sampling and selection process

Due to the qualitative approach of this study, a non-probability sampling technique was used. Specifically, purposive sampling technique was selected for the retrofitting projects. Zikmund ([32], p. 382) defines purposive sampling as "a non-probability sampling technique in which a researcher selects the sample based on his/her judgment about some appropriate characteristics required of the sample members". Thus, through a process of purposive sampling ([33], pp. 381–385) the researcher selected five case study buildings from the categories of retrofitting projects involving LCBs. The selected projects for this study were chosen from LCB retrofitting projects in the East of England. However, unlike statistical sampling, the sample is not a representative of the entire population of LCBs in England. Although the selected projects have various types of ownership, however, they are used for similar purposes.

3.2 Data collection process

The data were collected through phone interviews, site interviews and case study buildings. The researcher conducted some interviews with the building tenants to collect information on how energy is used in the building. Following the phone interview, the researcher conducted six on-site interviews with the building managers/operators to learn more about the management practices being implemented in their building. The purpose was to build on the phone interviews to obtain a more in-depth view of the operational performance of the building and to gain a deeper understanding of the best practices in operating the building. Energy use data collection formed the main focus of the data collection of the selected buildings. According to [34] annual energy use can be estimated either by using top-down approach or a bottom-up approach. The bottom-up approach involves the use of the calculation methods while the top-down approach involves an analysis of measured energy consumption and appropriating it to the elements responsible for energy consumption. The bottom-up approach is mainly founded on theory, the calculated loads, and the rated capacity of energy-using equipment. The limitations of the bottom-up approach lies in the weakness of the calculated results rarely agree with metered data. This leads to overestimating energy consumption and in masking individual elements of energy use. Thus, the top-down approach was preferred for this study because of its advantage of providing a greater degree of accuracy as it is based on factual metered data [35]. In the top-down approach, the actual measured data are obtained from utility companies such as monthly utility bills and meter prints outs and are critically examined. The rational for this is to estimate the annual energy consumption and to determine how energy is being used for the activity within the building. Utility data from the buildings were collected for 12 months and the figures were converted to kg of CO_2 and ranked in order of absolute energy consumption.

3.3 Data analysis method

The data analysis method for this study comprise of two approaches. Firstly, benchmarking was adopted as an energy performance tracking strategy. It is a strategy most often use in normalising energy consumption-based metrics such as weather or square footage to promote realistic comparisons with other similar buildings [36]. Benchmarking as used in the analysis of the data in this study is the most prevalent performance tracking approach found in the literature capable of providing a high level picture of energy use. CIBSE TM46 [37] energy benchmarks were adopted to benchmark the performance of the investigated buildings (**Table 2**).

Secondly, the ranking was adopted to categorise higher performing buildings from the lower performing ones. Although the review of literature indicates that numerous ranking and scoring systems have been developed, however, there is no scientific consensus method [38]. The use of ranking will enable the building owners and the facilities managers to be able to compare their building performance to similar building's size and similar pattern of use in order to be adequately informed on the actions to be taken to boost the performance of their buildings. Energy use of the surveyed buildings was converted into CO_2 emission using DEFRA [35] CO_2 emission conversion factors. It assumes CO_2 emission factors of 0.184 kg of CO_2/kWh for gas and 0.542 kg of CO_2/kWh for electricity. The carbon emissions of the buildings were calculated to determine both 'absolute' and 'relative' terms. The absolute emissions indicate the total footprint while relative emissions refer to the absolute figure indexed to a unit of this per m^2 per performance also referred to as 'intensity indicators'. During the analysis of data, the interpretation and the presentation of results; ethical issues were taken into consideration

	Benchmarks (kWh/m^2)	Units (m^2)	Benchmarked annual utility consumption (kWh)
Gas	105	390	40,950
Electricity	20	390	7800

Source: CIBSE TM46:2008 energy benchmarks.

Table 2.
Annual utility benchmarking.

by intentional coding the surveyed buildings using an alphabet (**Table 4**) to keep the building's identities and location hidden. This is in line with the suggestion of ([39], p. 89) that the process of data collection should not put participants at risks and that the vulnerable population should be respected by the researcher.

4. Results and discussions

A benchmark comparison of surveyed buildings was performed; first to provide an indication of how the buildings are performing; second, to identify where energy waste is prominent, and third to identify the areas for improvement. **Figure 3** shows the result of the comparison between the benchmark and annual energy consumption of the buildings surveyed.

It could be observed that the energy use of the buildings was substantially and simultaneously higher and plateaued than the benchmarked utility consumption, apart from buildings 'B1' which had lowest energy consumption (i.e. better than the benchmarked utility consumption). The energy performance indicator (EPI) for the investigated buildings is depicted in **Table 3**. It can be seen that the total annual energy use per heated floor area ranges from 17 to 730 kWh/m^2/year with a mean of 321.6 kWh/m^2/year. Building 'B1' was found to have the lowest EPI of 17 kWh/m^2 while building 'B5' was found to have the largest EPI of 730 kWh/m^2/year. Accordingly, the CO_2 emission from the buildings is shown in kgCO_2/m^2 per floor area in order of their emissions to allow for comparisons. **Table 4** shows the building characteristics and the pattern of use. It was observed that buildings used for catering services (B2, B3 and B5) recorded high energy use when compared to other uses. Meanwhile, building used for online bookshop (B1) recorded low energy usage per floor area. To facilitate comparison of energy use, according to the building pattern of use, total energy use in each category was determined and given

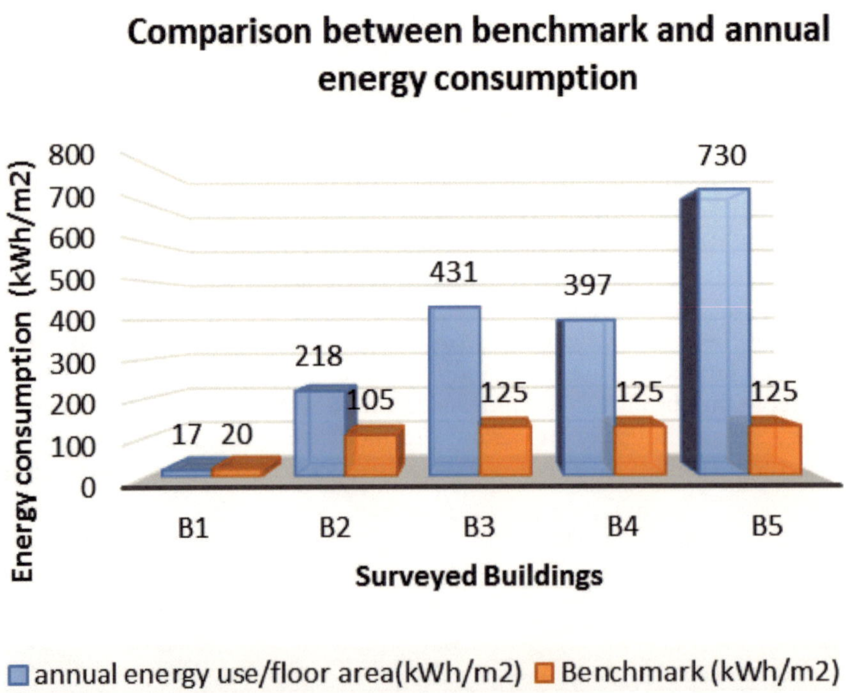

Figure 3.
Comparison between benchmark and annual energy consumption of the surveyed buildings.

	B1	B2	B3	B4	B5
Annual energy use (kWh)	4437	76,170	234,827	111,261	73,712
Annual energy use/floor area (kWh/m^2)	17	218	431	397	730
Annual CO_2 emissions/floor area (kgCO_2/m^2)	9	40	128	140	154
Energy performance category	High	Low	Low	Low	Low
Rank	1	2	3	4	5

Table 3.
Energy performance of surveyed buildings by ranking.

	B1	B2	B3	B4	B5
Built	C15	C15	C15	C14	C17
Grade listing	I	I	I	I	II
Main uses	Online bookshop	Community cafe	Community cafe	Bookshop	Community cafe
Secondary uses	Online bookshop	Community cafe	Worship	Community cafe	Community cafe
Floor area (m^2)	269	349	545	280	101
Type of energy use	Electricity	Gas	Electricity & gas	Electricity & gas	Electricity & gas

Table 4.
Building characteristics and pattern of use.

overall rank (**Table 3**) according to their performance range (1 = high performance, 5 = low performance). Building 'B1' ranked 1st and the best performing having the least environmental impact.

Building 'B2' used as community café ranked 2nd with energy use more than twice compared to the benchmark. Energy use became more than tripled with building 'B3' used for dual purpose (i.e. community café and worship) and ranked 3rd. Building 'B4' (397 kWh/m^2) used for dual purpose (i.e. bookshop and community café) ranked 4th. Meanwhile, building 'B5' with a singular use for community café use the largest amount of energy (730 kWh/m^2) ranked 5th as the lowest performing building. The fuel type used by the buildings was investigated; the operational energy performance of the building (B2) using only gas energy was poor compared to the building (B1) using electricity only given the similar construction properties of the buildings. To compare and contrast between the performances of the buildings (B2 and B5), the overriding element that makes building 'B5' less performing to building 'B2' could be attributed to the level of demand, its use requires compared to building 'B2' or put another way, how desirable is its use. Further, the building location factors and the energy efficiency characteristics of the building is a key feature that should not be overlooked but instead needs to be better understood. Meanwhile, the risk posed to a building having such a very poor energy efficiency performance could become undesirable with time and become unfit for purpose. Thus, it can be seen that buildings used as a community cafe either as single use type (B2 & B5) or used in combination with other functions (B3 & B4) appears to consume more energy when compared to other uses. Apart from space heating, the high energy use of these buildings is perceived to be as a result of multiple factors arising from energy end uses. For instance, process plant (e.g. freezers, fridges, etc.) and other equipment (e.g. catering), user's behaviour and attitude, efficiency of heating equipment, etc. It is estimated that around 25% of the energy used for catering operations is expended in the preparation, cooking and serving food. By far the largest proportion of this energy is consumed by cooking apparatus from which much of it is wasted through excessive use, poor utilisation and poor energy management attitude.

Further observation from **Table 4** show that as the building size (i.e. B3, B4 and B5) decreases, energy consumption increases. This finding is quite surprising and contrary to expectation that smaller size building (B5) would consume less energy. The increase in energy consumption in smaller size buildings could perhaps be attributed to the intensity of energy use and more patronage than the larger ones and operational practices of the building operators. In addition, the preference in the use of smaller buildings may have consequently resulted in their over-use, which could have also been responsible for their high energy consumption. Further findings show that among the investigated buildings, only building 'B1' had a form of energy management strategies apart from the fact that the pattern of use contributes to its low energy use. Generally, there are two methods to effectively reduce the energy demand of a building. The first and the most common approach is the physical improvement to the buildings (i.e. fabric and services). The second approach is to improve the way the building is operated (i.e. through facilities management and users behavioural change). However, the peculiarities of Heritage buildings (e.g. listed churches) such as their thick masonry walls, stained glass windows, traditional organic building materials, lime plasters/lime wash and the way they absorb and release moisture; pose challenges and limitation to modern applications of energy efficiency measures. Therefore, the first approach has limited application in several ways. For instance, application of modern type of insulation could create excessive humidity and dampness damaging the fabric irreversibly. Whilst a balance between air tightness and unwanted heat loss through the envelope and controlled ventilation needs to be found; the second approach, which is more passive would be more appropriate.

The most sustainable and available options for heritage buildings is to actively engage users and visitors in an energy saving campaign, introduce energy management systems and making building services such as heating and lighting more efficient. Public building users generally do not have incentives to act in an energy efficient manner. This is the case for all types of users. The result from this study reveals the need for energy management policies and strategies to minimise the energy required to operate rehabilitated Heritage buildings and to ensure their long term sustainability. This is due to their nature as 'hard to treat buildings'. Thus, it is this project's contention that the operational energy efficiency policy should be developed and implemented for sustainable retrofitting of heritage buildings in the world level and not only at EU level.

5. Recommendations and implications for sustainable practices

Existing buildings, particularly those of historical significance, can be transformed with a wide range of interventions, a process which greatly relies on the peculiarities of each case. However, the designer needs to assess what is best for the building and its future users/occupant. A secondary objective of the project should be to investigate and assess proposed functions of the new and upgraded building through the viewpoint of low energy use since energy use of a building is greatly affected by its use and the occupancy patterns that it creates as evidenced by this study. Thus, low energy use as a key contemporary demand for better performance and as a response to climate change should be fully integrated into the retrofitting of heritage buildings. In this way, energy use in rehabilitating heritage building projects can also provide insights for the selection of the appropriate future use, and whether that use can be a viable option for its operation. Further recommendations include considering the potential of integrating building management systems into any proposed retrofitting projects. This allows the monitoring and controlling the heating, cooling and lighting systems as well as ventilation systems where it is introduced in different parts of the building at different times of the day.

Generally, the traditional heritage building technology did not tightly seal buildings neither did they use damp-proof courses in walls nor damp-proof membranes below ground floors. This results in high ventilation rate within the building and consequently high energy consumption. Thus, achieving balance between moisture content, water vapour and ventilation within the building is achieved by high ventilation rates through operable windows, doors and different kind of gaps. By adapting historic buildings, this equilibrium is disrupted, as a rule, providing a building with an insulation envelope is accompanied by an increase in air-tightness, with the aim of reducing the heat losses and making the insulation work. In order to achieve a good energy performance, the available and the most inexpensive and effective option is draught proofing. Nevertheless, this disrupts natural ventilation, making a ventilation system essential to secure the necessary air exchange rate and to maintain the interior air quality. Meanwhile, where internal partitions are used, they can be linked to the function room booking system so that lighting, heating and cooling are only switched on when a function is going to be in use.

Other areas could be fitted with movement or occupancy sensors as part of a wider building management system, so lights come on only when people are present. Similarly, the use of daylight sensors can control artificial lighting according to what is required in different areas of the building, based on natural light entering the building from outside. Building management systems are considered more cost-effective for large Heritage buildings (e.g. churches) used for community and commercial purposes. Further, building owners and corporate building occupiers (i.e. users) and the professionals should be made aware that one of the overriding factors that make a sustainable building is the level of its reduced energy demand when occupied. Therefore, behavioural change of the users should be targeted by making real time information about energy use available. The energy behaviour of employees can also be influenced and changed by providing them with current information about their energy use at their desk, room and/or section within the building. Consumers would also need to be made to understand that lower energy running costs of the buildings means higher operating profits and less impact on the environment.

Additionally, the appointment of personnel trained in energy management as building operators for Heritage buildings retrofitting projects is imperative as this has been known to dramatically reduce energy consumption by 40% and consequently advanced improved operational energy performance of retrofitting projects. More importantly, after all the minimum intervention options have been exhausted for energy saving, consideration for generating on-site energy from renewable (e.g. air-source heat pumps, ground source heat pumps, biomass boilers, etc.) sources could also be sensitively installed on the buildings. Although this option could also be considered earlier where there is already a history of an on-site energy generation, or where boilers are being replaced. The professionals involved in heritage buildings retrofitting projects, such as architects, installation engineers, building surveyors etc., should include services such as analysis of whole life costs and carbon savings in services they provide to support the justification of the investment. Achieving the levels of improved energy performance in the retrofitting of heritage buildings required would not be likely reached if professionals rely on marketing only the economic benefits and payback periods to potential clients. The retrofitting projects should be seen as an opportunity to reduce long term expenditure on energy use by tackling the two simultaneously.

6. Conclusion

The purpose of this paper was to assess the current performance of the existing retrofitting of heritage buildings through the viewpoint of energy efficiency. Findings from the study support motivation for conducting this study as it shows that in terms of energy performance, the majority of retrofitting projects of heritage buildings are currently underperforming. This study has shown that the quantity of energy used in the retrofitting of heritage buildings depends upon how intensity the building is utilised and how the building is operated. Thus, to effectively deal with energy use in retrofitting of heritage buildings, it is necessary to understand the purpose for which the building is to be used, the energy use implications for the new use, the building characteristics and its energy using systems. In addition, adequate knowledge is required of the options available to improve energy utilisation, the techniques for modifying buildings and systems, and the feasibility of replacing portions of them. Thus, it is recommended that lower energy use should be a key consideration in determining the effectiveness of any proposed interventions to the retrofitting of heritage buildings. Energy consumption of the resulting interventions and the possibility of energy generation should be the means for evaluation of actions taken to rehabilitate heritage buildings. The wider benefits of improved energy performance in retrofitting projects apart from improved thermal comfort can become a strong contributor to sustainable retrofitting of heritage buildings. Although the challenging factors impacting sustainable retrofitting of heritage buildings are varied and total elimination of the factors is difficult. However, sustainable management strategies towards the minimisation of their energy use should be aimed at provided such ameliorating strategies would not contravene the conservation policies and the requirements for their protection. New technological approaches with measures for low energy use should be explored not just for their retrofitting and adaptation capability but also for their sustainability.

Acknowledgements

The author would like to thank and acknowledge the support of the Department of Engineering and the Built Environment, Anglia Ruskin University, Chelmsford, United Kingdom for the sponsorship of this research.

Author details

Oluwafemi K. Akande
Department of Architecture, Federal University of Technology, Minna, Nigeria

*Address all correspondence to: fkakande225@googlemail.com

References

[1] European Union. Directive 2009/28/EC European Parliament and of the Council on the Promotion of the use of energy from renewable sources. Available online: http://eur-ex.europa.eu/LexUriServ/LexUriServ.do?uri=Oj:L:2009:140:0016:0062:en:PDF. [Accessed: 25 March 2013]

[2] European Union. Directive 2010/31/EC of the European Parliament and of the Council on the energy performance of buildings. Available online: http://eur-lex.europa.eu/LexUriServ/LexUriServ.do?uri=OJ:L:2010:153:0013:0035:EN:PDF. [Accessed: 25 March 2013]

[3] European Parliament and the Council of the European Union. Directive 2012/27/EU on energy efficiency. Official Journal of the European Union L. 2012;**315**:1-56. Available online: http://eur-lex.europa.eu/JOHtml.do?uri=OJ:L:2012:315:SOM:EN:HTML [Accessed: 25 March 2013]

[4] UNEP (United Nations Environment Programme). Buildings and Climate Change Status, Challenges and Opportunities. UNEP Publications; 2007. pp. 1-87 https://issuu.com/aesmith/docs/buildings_and_climate_change

[5] UNEP (United Nations Environment Programme). Buildings and Climate Change: Summary for Decision-Makers. UNEP Publications; 2009. pp. 1-62 http://ledsgp.org/resource/buildings-climate-change-a-summary-for-decision-makers/?loclang=en_gb

[6] British Standards Institution (BSI) 7913. Guide to the Conservation of Historic Buildings. BSI Standards Publication; 2013

[7] Prudon THM. Preservation of Modern Architecture. USA: Wiley; 2008

[8] Feilden MB. Conservation of Historic Buildings. Oxford: Architectural Press; 2003

[9] Genre JL, Flourentzos F, Stockli T. Building refurbishment: Habitat upgrading. Energy and Buildings. 2000;**31**(2):155-157

[10] Pugh C. The costs and benefits of retrofitting and refurbishment. Property Management. 1991;**9**(2):143-156

[11] Kohler N. A European perspective on the Pearce report: Policy and research. Building Research and Information. 2006;**34**(3):287-294

[12] van Beuren E, de Jong J. Establishing sustainability: Policy successes and failures. Building Research and Information. 2007;**35**(5):543-556

[13] Bradley PE, Kohler N. Methodology for the survival analysis of urban building stocks. Building Research and Information. 2007;**35**(5):529-542

[14] Remoy HT, van der Voordt TJM. A new life: Conversion of vacant office buildings into housing. Facilities. 2007;**25**(3/4):88-103

[15] Velthius K, Spennemann D. The future of defunct religious buildings: Ditch approaches to their adaptive reuse. Cultural Trends. 2007;**16**(1):43-66

[16] Kurul E. A qualitative approach to exploring adaptive re-use processes. Facilities. 2007;**25**(13/14):554-570

[17] Langston C, Wong FKW, Hui ECM, Shen L-Y. Operational assessment of building adaptive reuse opportunities in Hong Kong. Building and Environment. 2007;**43**(10):1709-1718

[18] Latham D. Creative Re-Use of Buildings (Vols. 1-2). Shaftesbury, UK: Donhead Publishing; 2000

[19] Fitch JM. Historic Preservation: Curatorial Management of the Built

World. New York: McGraw-Hill Companies, Inc; 2001

[20] Rodwell D. Conservation and Sustainability in Historic Cities. Oxford, UK: Blackwell Publishing; 2007

[21] Langston CF, Wong EH, Shen LY. Strategic assessment of building adaptive reuse opportunities in Hong Kong. Building and Environment. 2008;**43**(10):1709-1718

[22] De Rohan-Chabot. Future for Religious Heritage. 2013. Available online: http://www.frh-europe.org

[23] Church Commissioners for England Annual Report, 2010. 2014. Available from: https://www.churchofengland.org/.../final%20annual%20report%20at%2019%20april

[24] English Heritage. Power of Place: The Future of the Historic Environment. London: English Heritage; 2001

[25] Parliamentary Assembly. Redundant religious buildings. Resolution 916 of the assembly debate (3rd sitting) 1989. Available at: http://assembly.coe.int/nw/xml/XRef/Xref-XML2HTML-en.asp?fileid=16327&lang=en

[26] Worthing D, Bond S. Managing Built Heritage: The Role of Cultural Significance. Oxford: Blackwell Publishing; 2008

[27] Bowitz E, Ibenholt K. Economic impacts of cultural heritage—Research and perspectives. Journal of Cultural Heritage. 2009;**10**:1-8

[28] De la Cruz-Lovera C, Perea-Moreno AJ, De la Cruz-Fernández JL, Alvarez-Bermejo JA. Worldwide research on energy efficiency and sustainability in public buildings. Sustainability. 2017;**9**:1294

[29] Polo López CS, Frontini F. Energy efficiency and renewable solar energy integration in heritage historic buildings. In: Energy Procedia. Vol. 48. 2014. pp. 1493-1502

[30] Akande OK. Factors influencing operational energy performance and refurbishment of UK listed church buildings: Towards a strategic management framework [unpublished PhD thesis]. Chelmsford, United Kingdom: Anglia Ruskin University; 2015

[31] Oppenheim AN. Questionnaire Design, Interviewing and Attitude Measurement. UK: Pinter Publishers; 1992

[32] Zikmund WG. Business Research Methods. 7th ed. Mason, OH: Thomson South-Western; 2003

[33] Tull D, Hawkins D. Marketing Research and Method. 2nd ed. New York: Macmillan Publishing Co; 1980. pp. 381-385

[34] Turner WC. Energy Management Handbook. New York: Wiley-Interscience; 1982

[35] Effinger M, Friedman H, Moser D. Building performance tracking in large commercial buildings: Tools and strategies. Online Sources: 2010. www.cacx.org/PIER/documents/Subtask_4-2_Report.pdf. [Accessed: 10 September 2015]

[36] Chartered Institution of Building Services Engineers (CIBSE). Energy Benchmarks: CIBSE TM46. London: CIBSE; 2008

[37] Davis GA, Swanson M, Jones S. Comparative Evaluation of Chemical Ranking and Scoring Methodologies. University of Tennessee, Centre for Clean Product and Clean Technologies; 1994. Available from: http://isse.utk.edu/ccp/pubs/pdfs/CECRSM.pdf [Accessed: 17 September 2015]

[38] DEFRA (Department for Environment Food and Rural Affairs). E-Digest of Statistics. 2009. Available from: http://www.defra.gov.uk/environment/statistics/globatmos/gakf07.htm

[39] Creswell JW. Research Design: Qualitative, Quantitative, and Mixed Method Approaches. 3rd ed. Thousand Oaks, CA: Sage; 2009

Chapter 3

New Additions to Existing Built Heritage and Their Contributions to Sustainable Development: Cases from Ankara, Turkey

Gulsen Disli

Abstract

New additions to historic buildings are mostly required either to extend the longevity of the building or to meet the new program requirements imposed to the built heritage as part of historic preservation process. The additions might be in the form of a rooftop, front, rear, side, or basement attachment. In all cases, the question of what is a sensitive addition according to the world-wide preservation standards is to be well analyzed. This study uses data from a new survey on five case study historic mosques and khans in Ankara with new exterior or interior additions to reveal the quality, compatibility and/or incompatibility of contemporary new additions and their contributions to sustainability. Case study analysis, in-situ observations, archival and literature survey are the principle methods applied during the study. Research findings show that, the additions follow different paths; they can be differentiated from the main historic building with their massing, material, and either color, or they hinder the existing built heritage, or even damage its character-defining features. Hence, for sensitive and successful new additions, restorations should be in compatible with world-wide standards and should be well analyzed and applied by the related authorities both during project approval and restoration phases.

Keywords: existing built heritage, historic preservation, rehabilitation, adaptive reuse, new additions, alterations, sustainability, Ankara

1. Introduction

Historic buildings bear the traces of different periods because of additions attached to the building in different times. Among them, the contemporary new additions, commonly applied as part of historic preservation treatment, either in the form of exterior additions or interior alterations, expand the outer limits of the original building, or create new spaces in building interiors [1]. A new exterior addition is mostly required for programmatic reasons and is accepted as an appropriate alternative only if "the new program requirements cannot be met within the existing building envelope" or by altering only interior spaces without changing character defining features of the building [1–4]. Interior additions are similarly applied to make compatible use of the property through interior repairs and alterations [4]. As US Standards for Rehabilitation & Guidelines for Rehabilitating Historic Buildings has proposed and if an exterior addition has to be done, it should be clearly differentiated from the original and should be built without damaging character defining features of the old [4]. This research examines the new exterior additions and interior alterations observed in five case study historic buildings, three khans and two mosques, across historic Ulus district, in Ankara. They provide examples of both adaptation projects and original use, all maintaining significant details of the old building, and at the same time have new additions that attract public attention. Selection parameters are; they should be public buildings, they are to be located in historic fabric of city of Ankara, and their new additions should exemplify either the need for adaptive reuse or continuous use of original function of the old building. Main objectives of the study are first to question the compatibility of the new additions and alterations with international standards and to investigate their contributions to sustainable development of the district. It begins with a literature review on the type of new additions to existing built heritage, followed by international legal framework and guidelines on new additions and continued with case study buildings and their new additions with an evaluation of their compatibility with the standards and their sustainable contributions.

There is a vast amount of literature on new additions to historic buildings [5–14]. Among them, there are various studies examining the issue of evaluating those new additions, possible approaches applied in their designs, and their appropriateness to the historic fabric [5–10]. New additions have also been discussed by scholars within adaptive reuse and rehabilitation framework and with a focus on sustainable benefits of rehabilitation [11–14].

There are also considerable literature related to prehistoric and historic development of historic city center of Ankara; on municipal, construction, agriculture, and commercial activities mainly in historic Ulus district, and related to urban, archeological, natural, and historical protected areas and strata's of the area [15–22]. Among them, Taş's research includes a comprehensive archival research on commercial activities, administrative structure, neighborhood, and public and private life in the seventeenth century Ankara [18]. Various studies similarly focus on the historic districts and architecture of the city including historic buildings belonging to Roman, Seljuk, Ottoman, and Republican periods [23–29]. Though in limited number, some scholars also examined the urban conservation process in Ulus district and alterations and special transformations of historical buildings in due course, as well as their sustainable conservation issues [30–33].

Although, all those existing literature tend to focus on architecture and history of Ulus district, and they rarely address the issues of new additions to historic buildings and their contributions to sustainable development in the area. Hence, this study is significant in terms of providing data on compatible and incompatible new additions to the case study old buildings located in the district and providing evaluations on their contributions to the revitalization of the area.

1.1 New additions to existing built heritage

New exterior additions, as part of rehabilitation treatment, can be inevitable in order to extend the longevity of the historic building falling into disuse because of the problem of abandonment, misuse, or damage. Changes in present day requirements, programmatic needs caused by adaptive reuse, and need for the completion of damaged parts in historic buildings esthetically, functionally, and structurally, as well as user expectations also cause differentiations in the form of exterior additions and/or interior alterations to the built heritage [9, 34]. Al-Jameel and Saffo (p. 3)

Figure 1.
New additions designed according to three different approaches: Standard, contrasting, and identical [3, 7, 39, 40].

also counts the "factors such as natural changes, disasters, social, economic, cultural and political transformations" among the very reasons of constructing new additions with different types and models [35].

There is not a certain formula about the type and form of the addition, such that it can be "traditionalist, contemporary or a simplified version of the historic building", as long as it preserves a balance between the differentiation and compatibility [36]. Though there are some recommendations, there are not certain rules on scale (height and width), form, massing, setback, orientation, alignment, rhythm, spacing, and proportion of a new addition [1, 3, 4, 37]. It can be as small as a vestibule, or as large as an entire building mass, but it is encouraged that it should be smaller than and not overwhelm the original building and should be less visible from the street side [2, 37]. Hence, it is more appropriate to position a new addition at the rear or side elevation of the existing building, and front additions are to be avoided as much as possible [37].

According to "The Secretary of the Interior's Standards for the Treatment of Historic Properties with Guidelines for Preserving, Rehabilitating, Restoring & Reconstructing Historic Buildings", White (p. 38) categorizes the various recommended and discouraged methodologies toward additions to historic buildings in three groups: standard, contrasting, and identical [7] (**Figure 1**). Standard approach is explained as "compatible," but "differentiated" from the old, thus ensuring that it "subordinates" to the historic building [7]. Also known as abstraction, in this approach, a new addition should be similar, but slightly different from the original building [6]. As another approach, contrasting style is defined as in "extreme contrast" to the old, thus avoids potential misunderstanding of what is original and what is new [6, 7]. On the other hand, in identical approach, new additions are almost the same with the old, in terms of style, design elements, material, scale, and detail [6, 7]. Semes similarly defines four possible strategies in designing new construction in a historic setting and calls them as the following; "(1) literal replication, (2) invention within the same or a related style, (3) abstract reference, and (4) intentional opposition [38]."

In addition to above mentioned design approaches, there are also some criteria that should bear in mind in designing new additions. Tanaç-Zeren (pp. 31–33) explains the criterion in new additions constituting the esthetic impression as follows [34]:

- "Environmental impact: the historic neighborhood in which the building is located and the building plot affects the location, material, and style of new addition.

- Impact of scale: horizontal and vertical dimensions of the new addition and its impact on human scale and its proportions affect the esthetic factors. Hence, new additions with correct dimensions/scales will not dominate the historic building.

- Impact of contrast: contrast can be set up by means of material, color, and scale.
- Impact of mass: selection of form in a new addition affects the sense of massing and builds its compatibility or contrast with the historic building.
- Impact of the rhythm: while designing new additions, it is possible to use rhythmic adaptation with the historic building by means of repeating the original material, proportion, or component.
- Impact of the material: material selection in new additions is an important criterion in determining quality of the addition."

According to the historic preservation standards, primary exterior new additions include [1, 3, 4, 7, 41] (**Figures 2** and **3**);

- Rear additions,
- Side additions,
- Camelback (second story) additions,
- Rooftop additions, and
- Storefronts.

Tanaç-Zeren (pp. 37–38) groups those new additions, which were generated as part of space and program requirements of secondary functions given to the old buildings, as follows [34]:

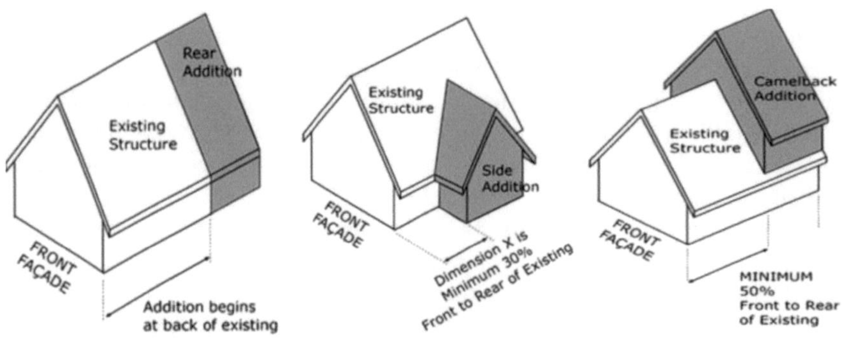

Figure 2.
Drawings showing different types of appropriate additions according to the City of Houston historic preservation manual, 2015 [7].

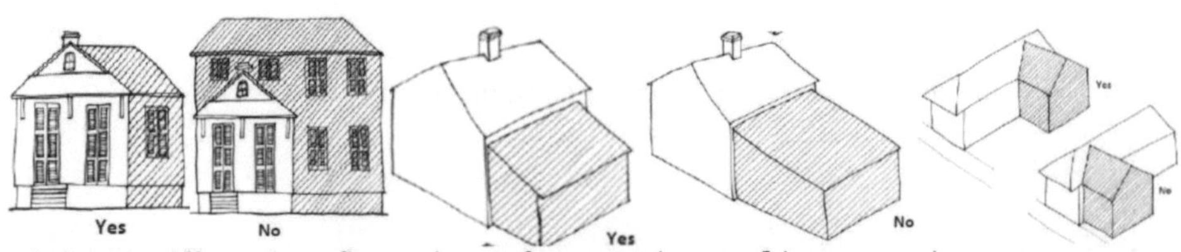

Figure 3.
Different configurations of appropriate and inappropriate new additions to historic buildings [37].

- "completions of roof,
- completions of façade,
- transition elements between the two buildings,
- fire escapes attached to the historic buildings, and
- eaves additions."

There are also front and basement additions observable in historic buildings. But, especially the front additions are mostly avoided or even prohibited in order not to hinder the visibility of the original building from the street side [37].

US Secretary of the Interior's Standards for Rehabilitation recommend that functions and services necessary for the new use are to be located in noncharacter defining interior spaces instead of constructing a new exterior addition [1]. Interior alterations may include "inserting an additional floor; an entirely new mechanical system; or creating an atrium or light well, but such alterations should not radically change, obscure, or destroy character defining spaces, materials, features, or finishes" [42].

2. International and legal framework and guidelines on new additions

In principle, world's leading preservation organizations such as United Nations Educational, Scientific and Cultural Organization (UNESCO), the International Council of Monuments and Sites (ICOMOS), and the U.S. National Park Service (NPS) oppose to the attachment of new additions to historic buildings [6, 10]. Nevertheless, they welcome compatible additions if required for the benefit of users and for present day use as long as they do not affect character defining features of the old building [6]. Those primary international charters, guidelines, and standards related to new additions to historic buildings and their related articles/sections are given below:

- Venice Charter, 1964, Article "12" and "13" [43],
- Third General Assembly of ICOMOS, 1972, Article "3" and "4" [44],
- The Declaration of Amsterdam, 1975, section "f" [45],
- U.S. Preservation Briefs 18, 1988 [46],
- U.S. Secretary of Interiors Standards for Rehabilitation and Guidelines for Rehabilitating Historic Buildings, 1990, "p. 62", Article "9" and "10" [4],
- U.S. Secretary of Interiors Standards for Rehabilitation and Illustrated Guidelines for Rehabilitating Historic Buildings, 1997, "pp. 91–93" [1],
- ICOMOS Charter of 14th General Assembly in Victoria Falls, Zimbabwe, 2003, Article "2.3" [47],
- UNESCO Fifteenth General Assembly, 2005, "p. 4", section "D-18" and "D-21" [48],

- ICOMOS New Zealand Charter 2010, "p. 8", Article "21" [49],
- U.S. Preservation Brief 14, 2010 [3],
- The Burra Charter, 2013, Articles "7.2., 15, 21.1, 21.2, 22.1" [50].

In addition to those international guidelines, in Turkey, there are also some regulations, which partially mention about the new additions to historic buildings though there is not a regulation solely related to the issue [51–53]. Among them, The High Council Resolution No: 731, dated 19.06. 2007, regulates the restoration interventions, applications, and inspections for the cultural heritage such as mosques, masjids, and tombs supervised and managed by Directorate General of Foundations [51]. According to this Resolution, in adaptive reuse of monumental historical buildings with waqf root, their original functions, if specified in their waqf documents, are to be preserved. Any kind of house reserved for someone who holds a particular job in a mosque or shops/stores cannot be built attached to the historic building itself or in their courtyards. Similarly, late comers' portico cannot be closed even with glass material, and in ablution spaces and in such areas, new material use that will destroy the traditional texture both structurally and visually is not allowed [51]. In addition, The High Council Resolution No: 660, dated 05.11. 1999, groups the immovable cultural heritage and regulates the conditions of their maintenance and repairs [52]. In this Resolution, it is noted that, if a new function is to be given to an old building, opinion of the Conservation Board is to be taken regarding the characteristics of new additions and their integration with the old, and those new additions can be applied only after its approval by the mentioned Conservation Board [52]. Furthermore, in The High Council Resolution No: 720, dated 04.10.2006, on Conservation and Usage Conditions in Urban Protected Areas, it is stated that it is not possible to build new buildings whose locations and height might adversely affect the protected urban silhouette of the street or zoning island that they are located [53]. Those existing regulations in Turkey show that they lack a detailed explanation about the approaches, types, and recommendations on application procedures of new additions to old buildings.

3. New additions to historic buildings and their contributions to sustainable development: cases from Ankara, Turkey

The case studies discussed in this part of the chapter are all historic public buildings, easily accessible located in historic Ulus district, in Ankara. The cases clearly exhibit both compatible and incompatible exterior and interior additions according to the international guidelines. Most of the examples selected were built in the sixteenth century, and the architectural landmark of the district and the additions were attached in last 80 years. In all of the cases, the exterior and/or interior additions, either in the form of small interventions or major additions, were attached to the existing building during the extensive restoration works. The case studies involve front, rear, side, basement, storefront, and rooftop exterior additions and some interior alterations. Among the case studies, the mosques are still used in their original function, and the khans exemplify adaptive reuse rehabilitation treatment; two are converted to a museum and one is started to be used as a boutique hotel. New exterior additions of case studies employ both standard and contrasting approaches explained in part 1.1. of this chapter.

3.1 Cases of historic buildings with compatible new additions

At this part, three cases—Çengel Khan, Çukur Khan, and Zağfiran Khan are selected which have contemporary additions, exemplifying both standard and contrasting design approaches. Original function of all cases is khan and located on the southwest side of historic Ankara Castle, one of the most important touristic routes of the district (**Figure 4**).

3.1.1 Çengel Khan

It is located in historic Ulus district, on the southwest side of the castle (**Figure 4**). According to its inscription panel, it was built in 1522–1523 by Rustem Pasha [28, 55]. It has an open-courtyard plan type constructed in masonry technique with stone and brick. Thanks to the sloppy topography of the site, two-story building has also a basement floor on the southwest side. The only entrance to the building is from the northeast façade, along which shops/stores are aligned. Its latest restoration was completed in 2005, and the building has started to be used as a museum with stores and a café, hence created new job opportunities. Adaptation of Çengel Khan to an exclusive cultural, social, and educational place attracts many local and international tourists. During the latest restoration, a steel and glass rooftop was added above the central courtyard, and thus, the courtyard could be used as an exhibition hall (**Figure 5**). Another new addition is observable on the basement floor, in the form of a rectangular room with large wooden windows at the southwest facade, used as a permanent exhibition area (**Figure 5**). On that facade, there is another rear addition located above the masonry walls (**Figure 5**). Similarly, wood and glass storefronts with wooden pillars and slightly sloppy awnings alongside the northeast façade of the building were added during the last restoration works (**Figure 5**). The masjid above the entrance vault changed its function, and this part was renovated with large wooden windows and started to be used as a café (**Figure 5**). Hence, in Çengel Khan, front, rear, rooftop, and storefront exterior additions are observable, all attached during 2003–2005 restoration works in order to meet the requirements of new function and usage of the building (**Figures 5** and **6**). Interior additions, on the other hand, are mostly seen in wet spaces and in the form of mechanical installations (**Figure 7**). Scale of interventions in both exterior and interior new additions is small, reversible, and differentiated with its material, form, and color, thus compatible with the old building.

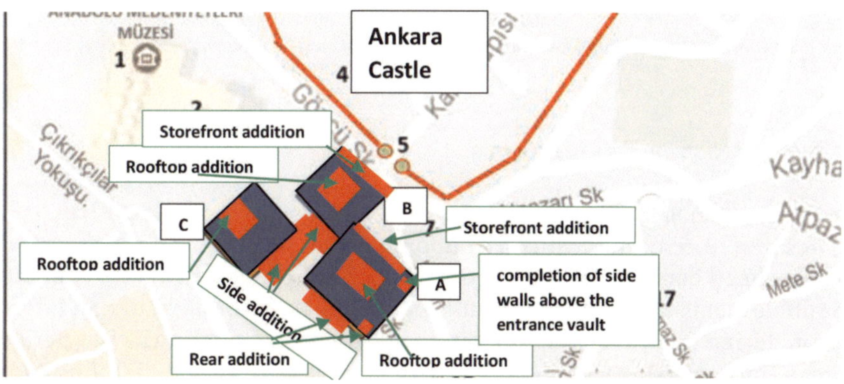

Figure 4.
A map showing the location of case study historic buildings with compatible new additions in ulus district. (A) Çengel khan, (B) Çukur khan, and (C) Zağfiran khan (Safran/Zafran) [54].

Figure 5.
Çengel khan: Rooftop addition covering the courtyard, exterior (a) [56] and interior views (b), completion of side walls above the entrance vault and storefront additions (c) and views from rear additions (d, e, and f) [39].

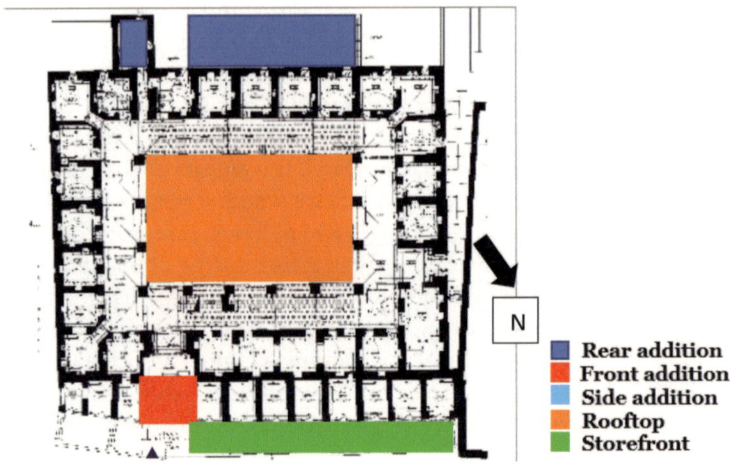

Figure 6.
Ground floor plan drawing of Çengel khan with new additions shown in different colors (source: Rearranged from [56]).

3.1.2 Zağfiran (Safran/Zafran) Khan

It is located in Salman Street behind (southwest side) the Çukur Khan. It was built in 1512 [28, 55]. The two-story building has an open central courtyard with the rooms lined beneath the semi-open riwaqs. Main body walls were constructed with rubble stone, and arches and vaults were built with brick material. Losing its function during the end of raw silk trade, it was used as a jail at the last periods of Ottoman Empire and at the beginning of the Republican Period [57]. Entrance to the building was from the west side [26]. As a result of new renovations and adaptations, today entrance is provided by means of a new addition passageway from the Cengel Khan. The building had many alterations in time and redesigned and adapted to a museum by Rahmi M. Koç Museum and Culture Foundation in 2012–2016. The upper level has been adapted for a restaurant with an open terrace. Regarding its contemporary exterior additions, Zağfiran Khan has a major side addition attached to its west facade and a rooftop addition above the central courtyard (**Figures 8–10**). New additions are clearly modern in their use of material, massing, and space arrangement. Rooftop architecture has a flat form built with glass and steel material (**Figure 8**). It creates a visual richness on the terrace of the restaurant and at the same time, enables permanent exhibition in the courtyard beneath. Side addition on the west side has similarly been created for various functional reasons, such as wet spaces, restaurant, and service halls for the stairs and elevators. Its facade is covered with glass material (**Figure 9**). Interior additions are mostly seen in west side addition, and in the old part, as door passageways and mechanical installations (**Figure 11**). In Zağfiran Khan, all those new additions exhibit present day appearance by means of contrasting style with

Figure 7.
Interior additions in Çengel khan are mostly observable at mechanical installations, handicapped lift, and door and window alterations and at designing wet spaces [39].

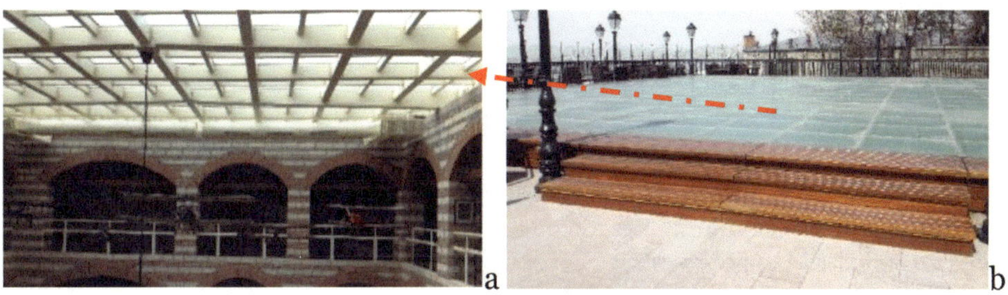

Figure 8.
Interior (a) and exterior (b) views of rooftop addition in Zağfiran Khan [39].

Figure 9.
Side addition of Zağfiran Khan; views from the restaurant, terrace (a), and whole south and west façade of the side addition (b) [39].

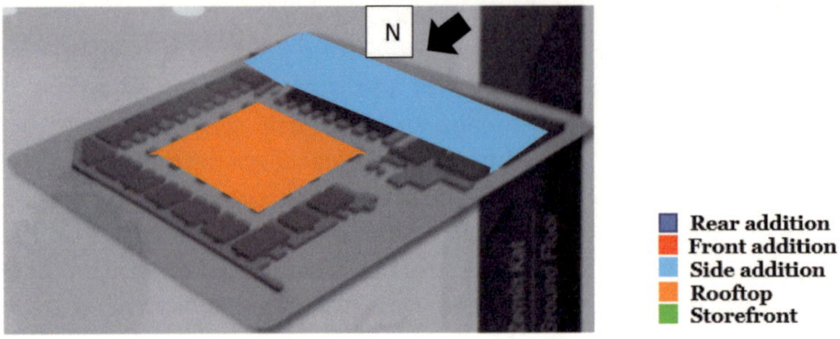

Figure 10.
Schematic plan drawing of Zağfiran Khan with new additions shown in different colors (source: Redrawn from [57]).

Figure 11.
Interior new additions of Zağfiran Khan; stairs, elevator, wet spaces, and newly added passageway between Çengel Han and Zağfiran Khan [39].

their material, color, and form, hence original and new parts can easily be differentiated from each other.

3.1.3 Çukur Khan

It was constructed in the sixteenth century with rubble stone foundation walls. Above the rubble stone foundation are wooden pillars and trusses with mud-brick infill. The two-story building has a wooden hipped roof. The building is located in Atpazarı, next to Çengel Khan. It has an open courtyard plan type, surrounded by semi-open riwaqs, and the rooms behind. Entrance to the building is provided on the north side [28]. It lost its historic character and character defining features to a great extent during its repair works in 1950s [26]. Hence, it was in a ruined condition before its last restoration which was completed in 2010. During the last adaptations, the building was converted to a boutique hotel with all the rooms having a different concept. Hence, it still retains its original accommodation use, and with the renovated shops on the north façade, it continues to provide public access/use and preserves its streetscape character. During its restoration in 2007–2010, its open courtyard was covered with a rooftop, and thus the space below could serve as a bar restaurant (**Figures 12** and **13**). This rooftop addition exhibits a contrasting approach in terms of material (glass and steel) selection and construction technique, so can easily be distinguishable from the original. The entire storefronts on the north facade, that were too deteriorated to repair or were not existent, were replaced with wood and glass material, by using the physical evidence as a model as suggested in The Secretary of the Interior's Standards for Rehabilitation and Guidelines for Rehabilitating Historic Buildings (**Figure 12**) [4]. Both the new rooftop and storefront additions are compatible with the size, scale, and massing of the historic

Figure 12.
Çukur khan: Rooftop addition (a and b) and storefronts (c) [39, 57].

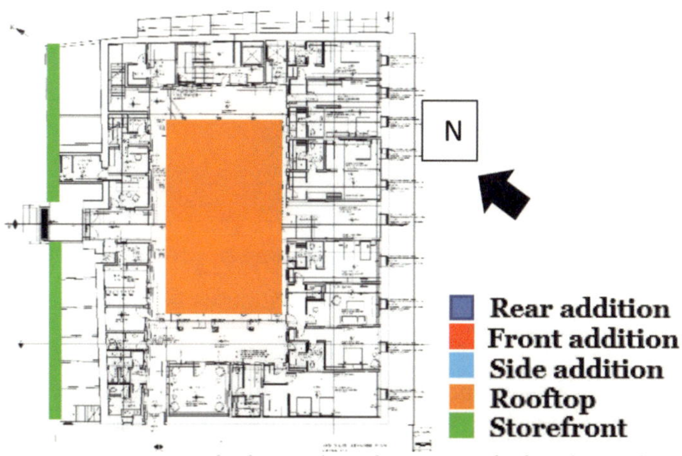

Figure 13.
Ground floor plan drawing of Çukur khan with new additions shown in different colors (source: Rearranged from [56]).

building. New interior additions include alterations and adaptations in wet spaces and mechanical installations.

3.2 Cases of historic buildings with incompatible new additions

At this part, two cases—Hacı Bayram-ı Veli Mosque and Kursunlu Mosque—are selected with exterior additions, exemplifying standard and contrasting design approaches. Original and present-day function of both cases is a mosque.

3.2.1 Hacı Bayram-ı Veli mosque

The building was constructed in 1427/1428 [26]. It is located in historic Ulus district, attached to August Temple of Roman period on the east side. Main prayer space has a rectangular plan, covered with a flat wooden ceiling, above which is a hipped roof with tile covering. Main body walls were built with brick material. On the south side of the mosque, there is Hacı Bayram-ı Veli Tomb, which has a square plan covered with a dome above [58]. The mosque undertook comprehensive repair works between 1703–1730 and 1757–1774, later in nineteenth century, in 1941, and in 1970 [26, 56, 58–60]. Especially during the last 80 years, the mosque was enlarged with new exterior additions, causing severe lost in its architectural and historic entity [58]. Late comers' portico on the north side of the mosque was enlarged, and a riwaq was added on the north, east, and west sides. Also, a basement floor was added for women's praying and for wet spaces. The first addition was conducted on the west side of the mosque with a vaulted gateway beneath, in early eighteenth century repairs [56]. Later, in 1941, late comers' portico was enlarged with a concrete addition covered with tile material [56]. It took its present-day appearance during its latest restorations completed in 2011 (**Figures 14** and **15**). At present, the front addition on the north side, together with its side riwaqs, is nearly larger than the original façade of the building. Though the old and new are at the same height, have similar material, and color, because of the location and massing of the new additions, the original building is suppressed by the new and cannot be perceived from the north side. Similarly with the basement addition, the square area

Figure 14.
Hacı Bayram-ı Veli mosque new exterior additions on the north side ((a) view from the north, (b) view from the east, and (c) view from the west) and views from the basement addition (d, e, and f) [39].

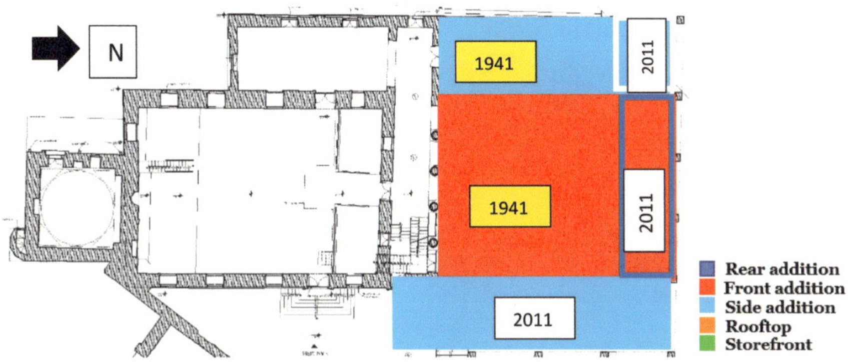

Figure 15.
Ground floor plan drawing of Hacı Bayram-ı Veli mosque with new additions shown in different colors (source: Rearranged from [56]).

Figure 16.
Exterior (a and b) and interior (c) views of front addition [39].

of the building was enlarged 2.5 times larger than the original size of the mosque itself [58]. Hence, in Hacı Bayram-ı Veli Mosque, exterior additions were applied with standard approach in the form of front and side additions, and basement addition exhibits the contrasting approach. They are all incompatible with the old building according to the international standards. On the other hand, regarding the interior alterations, providing access to the wet spaces and women's prayer space by means of a newly added escalator exemplifies the social responsibility role of new designs, taking into consideration of physically disabled and handicapped.

3.2.2 Kursunlu mosque

It is located in Ulus district, Samanpazarı area. Though the mosque does not have an inscription panel, it is dated to the sixteenth century [26, 55, 59]. It has a square main prayer area covered with a dome. Main body walls were built with alternating rows of stone and brick, and window arches and minaret were built with brick material. There is a minaret on the northeast side of the main prayer hall. The minaret was rebuilt after the earthquake in 1921 [26, 59]. The mosque was restored in 1914 and in 1990, and its late comers' portico on the north side was replaced with concrete two-story addition in 1972 repairs [28, 56] (**Figures 16** and **17**). Today, this front side addition is used for women's prayer area and for religious education of the children, but the original function of the main building is still continued. Its basement floor is allocated for wet spaces and water depot. The material used for the new addition—concrete columns and beams—is incompatible with the original building and also incompatible in terms of historic preservation rules. In addition, neither its massing and location nor its material is size respectful to the old building,

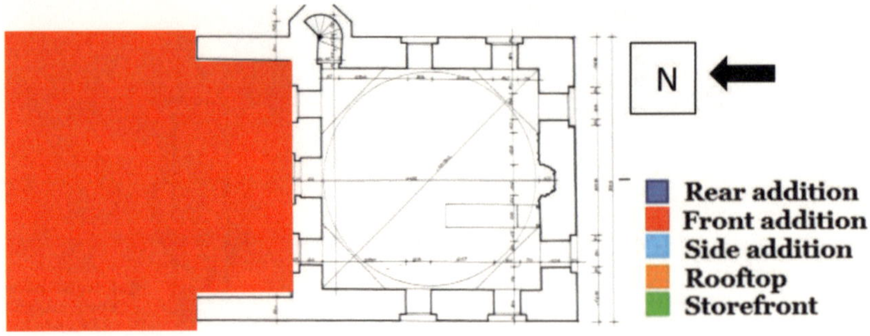

Figure 17.
Front addition in Kursunlu mosque located on the north side of the historic building (source: Rearranged from [56]).

and in that, the perception of old building is blocked from its north façade. Although, international guidelines suggest the control of mass, proportion, volume, material, color, and placing of new additions, in Kursunlu Mosque, and none of these criteria are observable.

3.3 Evaluation of case studies in terms of their new additions and their compatibility with the international preservation guidelines

As Torres (p. 6) states, in order to better evaluate the cohesiveness between historic buildings and their new additions, first, one has to clarify the methods of evaluation [6]. In order to establish an evaluation criteria and analysis method of compatibility of new additions, in their research, Yüceer and İpekoğlu (pp. 419–425) determined the architectural characteristics including "environment and setting, the site, the mass, and the facade order", and values of historic building before and after its new additions [9]. Mısırlısoy's (pp. 207–214) evaluation method, on the other hand, is based on the international guidelines [10]. Similarly, in this research, Mısırlısoy's (p. 213) evaluation chart has been developed more, and the five case studies of this research have been evaluated in terms of the type, approach, and compatibility of their new additions with the international conservation charters, standards, and guidelines on new additions to historic buildings that have been outlined in part two of this chapter. In this study, according to those international guidelines, total 11 subtitles have been determined as shown in **Table 1**. Then, the case studies have been assessed whether their contemporary additions meet or avoid a full level of cohesiveness with the old building according to those subtitles. In addition to the determination of evaluation criterion based on the international guidelines, archival and historical research, as well as field surveys constituted the other research methods. Among the five case studies, Çengel Khan, Hacı Bayram-ı Veli Mosque, and Kursunlu Mosque have front additions; Çengel Khan, Çukur Khan, and Zağfiran Khan have rooftop additions; Çengel Khan and Çukur Khan also have storefront additions; Hacı Bayram-ı Veli Mosque and Zağfiran Khan have side additions; Çengel Khan has a rear, and Hacı Bayram-ı Veli Mosque has a basement addition. Although in Çengel Khan, four different types of new additions—front, rear, rooftop, and storefront—are observable, in Kursunlu Mosque, only the front addition was applied (**Table 1**). In all cases, either the standard or contrasting design approaches were used, or for the compatible contrasting cases, new and modern materials such as steel and glass were incorporated. Regarding the cases which have rear, front, or side additions, the new attachments either maintain the same

Name and no of case study buildings	**Case study buildings**				
	Compatible			**Incompatible**	
	Compatibility of case study buildings				
1-Cengel Khan, **2**-Cukur Khan, **3**-Zağfiran, Khan, **4**-Hacı Bayram-ı Veli Mosque, and **5**-Kursunlu Mosque	1	2	3	4	5
Types of exterior additions: front (**a**), rear (**b**), side (**c**), basement (**d**), rooftop (**e**), and storefront (**f**)	**a, b, e, f**	**e, f**	**c, e**	**a, c, d**	**a**
Approaches in new additions: Contrasting (**C**), Standard (**S**), Identical (**I**)	**a-(C), b-(C), e-(C) f-(S)**	**e-(C) f-(S)**	c-(C) e-(C)	a-(S), c-(S) d-(C)	a-(C)
1-New additions preserve and respect the character and integrity of historic building	√	√	√	x	x
2- Reversible new additions	√	√	√, x	x	x
3- New additions can be differentiated form the old to preserve its character, by taking its design cues from, but not copy the historic building.	√	√	√	√, x	√, x
4- New additions can be compatible with the old with volume/mass, form, size, scale, color, texture, surface articulation, and material	√	√	√	√, x	x
5- Placement or location of new additions are carefully considered	√	√	√, x	x	x
6- New additions are simple and unobtrusive	√	√	√, x	x	x
7- New additions have less impact and are not highly visible from the public, and subordinate the historic building in size & design	√	√	√, x	x	x
8- New additions do not include imitations that falsify the public right of way	√	√	√	x	√
9- New additions are respectful and in harmony with the close neighborhood	√	√	√	√, x	x
10- New addition preserve the historic building's form/ envelope, significant materials and features	√	√	√	x	x
11- Rooftop additions are not much visible, have less impact, and inconspicuous when viewed from surrounding streets	√	√	√	—	—

Table 1.
An evaluation of case studies in terms of their new additions and their compatibility with the international preservation standards, charters, and guidelines.

height or less than the original building height, and general forms of the additions are rectangular cubes.

When evaluated according to the subtitles determined in **Table 1**, in terms of international standards, new additions in Çengel Khan and Çukur a Khan are the most compatible cases, and Zağfiran Khan similarly exemplifies a compatible case with their respectful design approaches both with the old building and with the neighborhood as well. On the other hand, Kursunlu Mosque and Hacı Bayram-ı Veli Mosque fulfill only two or three steps among the 11 subtitles, and thus have been categorized as incompatible cases. Their huge front additions, neither respect to the massing nor to the scale and proportion of the old building, at the same time dominate the historic fabric. In Kursunlu Mosque, the concrete attachment on the north side is also incompatible in terms of material selection.

4. Contributions of cases to sustainable development and rehabilitation of the building and the nearby built environment

Expanding damage to historic buildings because of natural changes, disasters, social, economic, cultural, and even political transformations, as well as the growth of the functional requirements and user expectations make the rehabilitation of old buildings in the form of new additions and adaptive reuse critical [35]. Thus, as Torres (p. 6) states, by reusing existing buildings, it becomes both possible to preserve the history and natural resources as well, by providing historic, economic, and sustainable contribution [6]. Kóródy and Vukoszávlyev, also, emphasize the importance of building rehabilitations in environmental, economic, and social responsibility, such that with their fewer new additions, and without demolition, rehabilitation projects both save energy, improve functional use, and "contribute to the local community's consciousness" [14]. Similarly, The Declaration of Amsterdam (1975) states that rehabilitation of old monuments is less costly than new construction and "social costs" are to be taken into account while deciding which one to choose [45]." Yüceer (p. 90) explains the economic value as "the potential of building for the satisfaction of maintenance expenses and for extra income with the financial gain in its present situation [8]."

When our cases are evaluated in that context, Cengel Khan, Cukur Khan, and Zağfiran Khan clearly exhibit economic, environmental, and social value with their new compatible additions and income generating new functions. Adaptation of those former khan buildings to museums and a boutique hotel with shops, café, and restaurant facilities increased the financial value of the buildings, and at the same time, both initiated a fine socio-cultural precinct in this part of Ulus district and brought new job opportunities back to this once intensively used buildings. Those cases, with their new use also demonstrate the importance of functional change and need for modernization for sustainability, as well as reveal the role of heritage conservation in urban revitalization. Such that they became important meeting and visiting points in the area attracting both local and international tourists, thus positively increased the quality of the vicinity and contributed to the economic, cultural, and social development of the local community.

Among the five case study buildings of this research, Hacı Bayram-ı Veli Mosque and Kursunlu Mosque still survive their functional continuity, though found incompatible according to international standards in terms of their new additions. Cengel Khan, Cukur Khan, and Zağfiran Khan were given a new function in order to meet the requirements of contemporary conditions with new exterior additions and interior alterations. In all cases, they positively contribute to the historical continuity and thus to the sustainability of heritage buildings, and urban fabric since they are still alive and can be transferred to next generations. Especially the compatible cases demonstrate that adaptive reuse of historic buildings can make them sustainable places linking the past with the present, as well as preserving their historic character. Also, considering the construction waste and cost, adaptation of historic buildings into new uses or retaining their original uses is also important in terms of economic and environmental point of view [61]. In all cases of this research, since they are still used in their original function or with their new functions by the attachment of new additions, instead of constructing new ones, construction wastes are reduced, and sustainability is increased. In addition, escalator, elevator, and handicapped lift attachments during their rehabilitations gave way to accessibility of physically disabled as part of social responsibility of sustainability. Hence, the cases presented in this research justify the importance of compatible new additions and adaptive reuse and at the same time, exemplify their contributions to sustainable development.

5. Discussion and conclusion

This research aimed to reveal the quality, compatibility, and/or incompatibility of contemporary new additions and their contributions to sustainability in five case study historic buildings in Ankara, Ulus district. In those cases, it has been determined that either standard or contrasting approaches have been used for the new additions. Compatible case study examples showed that thanks to those new additions and adaptive reuse treatments, and it became possible to contribute to the cultural, social, functional, and economic survivability of the historic buildings. But, it should be noted that since the human needs are consistent and unlimited, it is still possible that those buildings may be subjected to new additions in time that might cause to overwhelm the original building. In those circumstances, it should be strictly avoided to make again new additions to historic buildings. For instance, in Hacı Bayram-ı Veli Mosque and Kursunlu Mosque, it might have been a better solution to direct the users of the mosques to nearby ones, which are quite near to those buildings, during the prayer times, instead of constructing incompatible, overwhelming new additions. As Yüceer (p. 110) mentions, although refunctioning of historic buildings with new additions have some economic and functional gains, may at the same time cause lost "in the cultural, contextual and authentic values" in old buildings [8]. Hence, though new additions are required for sustainable development, the borders of new additions are to be determined thoroughly, and during their applications, both national and international standards and guidelines are to be obeyed strictly. Yet, when the local guidelines are examined, it is understood that, in terms of new additions, they are formulated at a general level, without direct information on types, possible approaches, and conditions of new additions, instead emphasizing that at specific situations, Conservation Boards are authorized and shall make a decision. In this research, especially the incompatible cases showed that, during the applications of new additions, international standards were not followed.

As future suggestions, this research can be a starting point for the evaluation of other historic buildings in Turkey with new additions, and it might give an insight for the new additions planned to be built to existing built heritage in the near future. It is also suggested to develop general local guidelines and standards on the issue, as the future suggestion. It is thought that increase in fieldworks by examining much more examples with new additions to historic buildings will contribute both to development of the local general guidelines and evaluation of their contributions to sustainability.

Acknowledgements

The author would like to thank Turkish Republic Directorate General of Foundations (DGF) for the provision of files on case study buildings.

Author details

Gulsen Disli
Faculty of Engineering and Architecture, Department of Architecture, Necmettin Erbakan University, Konya, Turkey

*Address all correspondence to: disli001@umn.edu

References

[1] Brown Morton W, Hume GL, Weeks KD, Jandl WH. Secretary of the Interior's Standards for Rehabilitation, and Illustrated Guidelines for Rehabilitating Historic Buildings. Washington, D.C.: U.S. Department of the Interior National Park Service Heritage Preservation Services; 1997. 135 p, here pp. 42-46, 90-93

[2] Baghasingh K. Sensitive Additions to Historic Buildings: A Primer [Internet]. 2015. Available from: https://www.heritageohio.org/2015/05/18/ sensitive-additions-to-historic-buildings-a-primer. [Accessed: 2018-10-22]

[3] Grimmer AE, Weeks KD. New exterior additions to historic buildings: Preservation concerns. Preservation Briefs. 2010;**14**:1-16

[4] U.S. The Secretary of the Interior's Standards for Rehabilitation and Guidelines for Rehabilitating Historic Buildings. Washington, D.C.: U.S. Department of the Interior, National Park Service, Preservation Assistance Division; 1990. 55 p

[5] Demel S. Preservation, historic significance and a theory of architectural additions: The canon and its consequences [thesis]. New York: Columbia University; 1996

[6] Guzmán Torres ZN. Historic buildings and contemporary additions: the elements of a cohesive design relationship [thesis]. College Park, Maryland: University of Maryland; 2009, here pp. pp. 6, 12-14, 20-21

[7] White AM. Defining appropriateness: An analysis of the Secretary of the Interior's Standards on additions to historic structures and its impact on the practice of architecture, development and preservation [thesis]. New York: Columbia University; 2018, here pp. 27-30, 38, 40, 42

[8] Yüceer H. An evaluation of interventions in architectural conservation: new exterior additions to historic buildings [thesis]. Izmir: Izmir Institute of Technology; 2005, here pp. 89-90, 110

[9] Yüceer H, İpekoğlu B. An architectural assessment method for new exterior additions to historic buildings. Journal of Cultural Heritage. 2012;**13**:419-425

[10] Mısırlısoy D. New designs in historic context: Starchitecture vs architectural conservation principles. Civil Engineering and Architecture. 2017;**5**(6):207-214

[11] Cramer J, Breitling S. Architecture in Existing Fabric: Planning Design Building. Basel, Switzerland: Birkhauser Verlag AG; 2007. 221 p. ISBN-13: 978-3038210955

[12] Thiebaut P. Old Buildings Looking for New Use: 64 Examples from Europe. Kustantaja: National Book Network; 2007. 276 p. ISBN-13: 978-3936681086

[13] Artola I, Rademaekers K, Williams R, Yearwood J. Boosting Building Renovation: What potential and value for Europe? Brussels: European Union; 2016. 72 p. Available from: http://www.europarl.europa.eu/RegData/etudes/STUD/2016/ 587326/IPOL_STU(2016)587326_EN.pdf [Accessed: 2018-10-15]

[14] Kóródy A, Vukoszávlyev Z. Aspects of social sustainability in cases of building rehabilitations in Central Europe. Építés – Építészettudomány. 2014;**42**(3–4):287-333

[15] TMMOB Mimarlar Odası Ankara Şubesi. Ankara Tarihi Kent Merkezi-1,

Ulus Paneli. Ankara: Matsa Yayıncılık; 2007. 139 p. ISBN: 978–9944893688

[16] Alpan A. Approaching the stratified city of Ankara as an integral unity. Idealkent Journal. 2016;**19**(7):662-671

[17] Galanti A. Ankara Tarihi I-II. Ankara: Çağlay Yayınları; 2005. 303 p. ISBN: 975-9899450

[18] Taş H. Ankara'nın Bütüncül Tarihine Katkı, XVII. Yüzyılda Ankara. Ankara: Türk Tarih Kurumu Yayınları; 2006. 281 p. ISBN: 975-1618711

[19] Koltuk T, Güneş M, Pakırdağ N. In: Alikılıç D, editor. Osmanlı Belgelerinde Ankara, Belediye ve İmar Faaliyetleri. İstanbul: Seçil Ofset; 2016. 451 p. ISBN: 9786059347112

[20] Güneş M, Koltuk N, Koltuk T, Pakırdağ N. In: Alikılıç D, editor. Osmanlı Belgelerinde Ankara, Tarım ve Ticaret. İstanbul: Seçil Ofset; 2016. 455 p. ISBN: 978-6059347150

[21] Kırpık G, Erdoğan A, Akyol H, Kılcı A, Çam M. Şehr-i Kadim Ankara-I. Ankara: Ankara Büyükşehir Belediyesi Yayınları; 2015 344 p. ISBN: 9786054067930

[22] Aydın S, Emiroğlu K, Türkoğlu Ö, Özsoy ED. Küçük Asya'nın Bin Yüzü. Ankara: Dost Yayınları; 2005. 720 p. ISBN: 9752981739

[23] Irklı-Eryıldız D, editor. Cumhuriyet Döneminde Mimari. Ankara: TMMOB Mimarlar Odası Ankara Şubesi Yayınları; 1999

[24] Kırpık G, Erdoğan A, Çam M, Kılcı A, Topraklı AY. Şehr-i Kadim Ankara-III. Ankara: Ankara Büyükşehir Belediyesi Yayınları; 2015. 494 p. ISBN: 9786054067954

[25] Konyalı İH. Ankara Camileri. Ankara: VGM; 1978. 117 p

[26] Öney G. Ankara'da Türk Devri Yapıları. Ankara: Ankara Üniversitesi Basımevi; 1971. 390 p., here pp. 59, 66, 67, 139, 141

[27] Öz N, editor. Ankara'nın Türk-İslam Dönemi Paneli. 14 October 2006. Ankara: TMMOB Mimarlar Odası Ankara Şubesi; 2008. 138 p. ISBN: 9759944896030

[28] Erdoğan A, Günel G, Kılcı A. Osmanlı'da Ankara. Ankara: Ankara Büyükşehir Belediyesi; 2007. 320 p. ISBN: 9799944473065, here pp. 119-122, 168

[29] Günel G, Kılcı A. 1924 map of Ankara city: Recognizing Ankara with an old map. Journal of Ankara Studies. 2015;**3**(1):78-104

[30] Aksoy Z, Varol Ç. Spatial transformation processes of immovable pious foundation properties from the early republic to today: Ankara case. Vakıflar Dergisi. 2015;**44**:105-129

[31] Erder E. Sustainable conservation issues of four 14th and 15th century mosques in Ankara: Ahi Elvan mosque, Örtmeli, Sabunî and Poyraji Mesjids. METU Journal of Faculty of Architecture. 2010;**27**(1):45-66

[32] Kurtar C, Somuncu M. Conservation and sustainability of urban cultural heritage: Ankara Hamamönü example. Journal of Ankara Studies. 2013;**1**(2): 35-47

[33] Osmançavusoglu A. Urban transformation process: Ulus historical city center planning project [thesis]. Ankara: Middle East Technical University; 2006

[34] Tanac-Zeren M. Tarihi Çevrede Yeni Ek ve Yeni Yapı Olgusu, Çağdaş Yaklaşım Örnekleri. İstanbul: Yalın Yayıncılık; 2010. 94 p. ISBN 9944313773, here pp. 29, 31-33, 37-38

[35] Al-Jameel AH, Saffo BH. Patterns of additions to old buildings. In: The Second International Conference on Conservation of Architecture, Urban Areas, Nature & Landscape: Towards a Sustainable Survival of Cultural Landscape- Heritage 2011; March 2011; Amman, Jordan. 2011. pp. 1-16

[36] Technical Preservation Services, Planning Successful Rehabilitation Projects, New Additions to Historic Buildings [Internet]. Available from: https://www.nps.gov/tps/standards/applying-rehabilitation/successful-rehab/additions.htm [Accessed: 2018-10-17]

[37] City of New Orleans HDLC – Guidelines for New Construction, Additions and Demolition [Internet]. 2011. Available from: https://www.nola.gov/nola/media/HDLC/Guidelines/12-New-Construction-Additions-Demolition.pdf [Accessed: 2018-10-15]

[38] Semes SW. "Differentiated" and "compatible": Four strategies for additions to historic settings. Forum Journal & Forum Focus. 2007;**21**(4). https://forum.savingplaces.org/viewdocument/differentiated-and-compatible

[39] Photo Archives of the Author, 2005, 2016, 2018

[40] Dinkeloo R. The Jewish Museum [Internet]. Available from: http://www.krjda.com/Sites/JewishMuseumInfo1.html [Accessed: 2018-10-18]

[41] Jandl HW. Rehabilitating historic storefronts. Preservation Briefs. 1982;**11**: 1-11

[42] Weeks KD, Grimmer AE. The Secretary of the Interior's Standards for the Treatment of Historic Properties with Guidelines for Preserving, Rehabilitating Restoring & Reconstructing Historic Buildings. Washington, D.C.: United States, National Park Service; 1995. 196 p. ISBN-13: 978-0160480614

[43] International Charter for the Conservation and Restoration of Monuments and Sites (The Venice Charter 1964) [Internet]. 1964. Available from: https://www.icomos.org/charters/venice_e.pdf [Accessed: 2018-10-21]

[44] Third General Assembly of ICOMOS Budapest, Resolutions of the Symposium on the Introduction of Contemporary Architecture into Ancient Groups of Buildings [Internet]. 1972. Available from: https://www.icomos.org/publications/93towns7e.pdf [Accessed: 2018-10-19]

[45] The Declaration of Amsterdam – 1975 Congress On The European Architectural Heritage [Internet]. 1975. Available from: http://www.icomos.org.tr/ Dosyalar/ICOMOSTR_0899129001496825552.pdf [Accessed: 2018-10-16]

[46] Jandl HW. Rehabilitating interiors in historic buildings identifying and preserving character defining elements. Preservation Briefs. 1988;**18**:1-8

[47] ICOMOS Charter- Principles for the Analysis, Conservation and Structural Restoration of Architectural Heritage, 14th General Assembly in Victoria Falls, Zimbabwe [Internet]. 2003. Available from: https://www.icomos.org/charters/structures_e.pdf [Accessed: 2018-11-01]

[48] UNESCO Fifteenth General Assembly of States Parties to the Convention Concerning the Protection of the World Cultural and Natural Heritage, Paris, 2005, [Internet]. 2005. Available from: whc.unesco.org/document/6814 [Accessed: 2018-10-20]

[49] ICOMOS New Zealand Charter for the Conservation of Places of Cultural

Heritage Value [Internet]. 2010. Available from: https://www.icomos.org/images/DOCUMENTS/Charters/ICOMOS_NZ_Charter_2010_FINAL_11_Oct_2010.pdf [Accessed: 2018-10-19]

[50] The Burra Charter, the Australia ICOMOS Charter for Places of Cultural Significance [Internet]. 2013. Available from: https://australia.icomos.org/wp-content/uploads/ The-Burra-Charter-2013-Adopted-31.10.2013.pdf [Accessed: 2018-10-17]

[51] Kültür ve Tabiat Varliklarini Koruma Yüksek Kurulunun Vakiflar Genel Müdürlüğü İdare Ve Denetiminde Olan Cami, Mescit, Türbe vb. Kültür Varliklarini Müdahale Biçimleri Uygulama Ve Denetimi İle İlgili İlke Karari [Internet]. 2007. Available from: http://teftis.kulturturizm.gov.tr/yazdir?549C7059D7FF142A3CF3B98AB092CB66 [Accessed: 2018-11-01]

[52] İlke Kararı- Karar No: 660 / Karar Tarihi: 05.11.1999 [Internet]. 1999. Available from: http://www.kulturvarliklari.gov.tr/TR-44311/ilke-karari–karar-no-660–karar-tarihi-05111999.html [Accessed: 2018-11-02]

[53] İlke Kararı- Karar No: 720 / Karar Tarihi: 04.10.2006 [Internet]. 2006. Available from: http://www.kulturvarliklari.gov.tr/TR-44332/ilke-karari–karar-no-720–karar-tarihi-04102006.html [Accessed: 2018-10-12]

[54] Ankaramızı Tanıyalım [Internet]. 2018. Available from: https://alivedatoygur.wordpress.com/2018/06/09/rahmi-m-koc-muzesi/[Accessed: 2018-10-18]

[55] Ergenç Ö. XVI. yüzyılda Ankara ve Konya. 1st ed. Ankara: Ankara Enstitüsü Vakfı; 1995. 247 p., here pp. 18-19, 24

[56] Archives of Directorate General of Foundations, file on Çengel Khan, Çukur Khan, Zağfiran Khan, Hacı Bayram-ı Veli Mosque, and Kurşunlu Mosque

[57] Rahmi M. Koç Museum Brochure. Ankara; 2018

[58] Eskici B. Ankara Hacı Bayram Camisi onarımlaru üzerine. Turkish Studies - International Periodical For The Languages, Literature and History of Turkish or Turkic. 2014;**9**(10): 557-576

[59] Erken S. Türkiye'de Vakıf Abideler ve Eski Eserler I-IV. 2nd ed. Vol. I. Ankara: Vakıflar Genel Müdürlüğü Yayınları; 1983. 768 p., here pp. 364, 417-418

[60] Ayverdi EH. Çelebi ve II. Sultan Murad Devri, 806–855 (1403–1451). İstanbul: İstanbul Fetih Cemiyeti Yayınları; 1972. 606 p., here p. 235

[61] Stewart E, Howells T, editors. New Uses for Heritage Places, Guidelines for the Adaptation of Historic Buildings and Sites. New South Wales: NSW Department of Planning; 2008. 52 p. ISBN 1 921121 130

Section 3

Vernacular Architectural Identity within Sustainability

Chapter 4

Sustainability and Vernacular Architecture: Rethinking What Identity Is

Maha Salman

Abstract

Sustainability has often been a fundamental part of the composition of both tangible and intangible cultural resources; sustainability and preservation of cultural identity are complementary. Elements of sustainable design are integral to vernacular architecture that have evolved over time using local materials and technology emerging from ambient natural and cultural environment creating optimum relationships between people and their place. This chapter aims to redefine what identity is as a concept and the impact of globalization on contemporary architecture especially on regions with rich heritage and unique culture as the Arab World. To accomplish this, the chapter examines the emergence of "local identity" as a reaction to the globalization of cultural values, uniform architectural styles, and stereotype patterns through discussing sustainability as a motivation for identity in culture and architecture. The research methodology is based on conducting a qualitative analysis of literature review to the main concepts discussed in this chapter such as: identity, culture, vernacular architecture, and sustainability. Through comparative analysis, the chapter investigates sustainability potential of vernacular architecture in the region to derive core concepts as guidelines of reproducing the characteristics of society and reveal identity of contemporary architecture in the Arab World.

Keywords: sustainability, identity, vernacular architecture, culture, heritage, globalization, contemporary architecture, Arab World

1. Introduction

Through history, human civilization has evolved and kept its continuation through integration with the surrounding environment and is dependent on the preservation of nature. Over time and the multifaceted interaction of evolution and human adaptability to the ambient environment, every place and region developed unique characteristics that distinguished it from other places, that is, the core of "identity."

Preserving the identity and special characteristics of a place requires in-depth understanding of the natural systems in place and immersion into the time-tested cultural responses to that environment's assets and liabilities, which contain the essence of sustainability.

The Arab World is one of the regions that embraces a diverse and rich architectural heritage; however, since the second half of the twentieth century, the vernacular architecture, in the Arab World, was exposed to extensive deterioration and destruction, due to the rapid modernization and drastic economic, social and cultural changes that took place in the region (**Figure 1**).

This uncontrolled expansion transformed major cities in the region into metropolitan areas and changed the traditional life style into a modern one. All these factors and drastic changes led to mislay the identity of the Arab architecture.

This chapter explores identity as an essence of architecture that goes beyond formal interpretation and visual metaphors through investigating sustainability potential of vernacular architecture in the Arab World.

Figure 1.
Modern high-rise buildings in Arab cities (Source: Salman, Maha).

2. Identity: people, culture, place, and architecture

In recent years, the issue of cultural identity in contemporary architecture has been attached a pivotal role in creating uniqueness and local identity in a competitive environment at a global level.

DOI: http://dx.doi.org/10.5772/intechopen.82025

The term “identity,” as defined in the Oxford English Living Dictionary, is the fact of being who or what a person or thing is; the characteristics determining who or what a person or thing is [1]. In Cambridge Dictionary, identity establishes or indicates who or what (someone or something) is, who a person is, or the qualities of a person or group that make them different from others [2]. In other words, identity means being unique and distinguished from others, and this can be applied to a thing, person, and group of people, society, country, or even nation.

Several natural and human factors contribute in defining “identity” such as place (region, geography, topography, and climate), people (society, community), and culture (traditions, customs, language, religion, and artifacts). It is essential to discuss these factors and how they are related to each other and how this was reflected in architecture.

Culture is one of the major factors that defines identity as it is related to people that created this culture. Vibhavari Jani, in her edited book “Diversity in Design: Perspectives from the Non-Western World,” suggests that culture refers to: “... distinctive way of life that represents values, customs, and norms of a group of people who pass these traditional values from one generation to the next. This learned way of life then reflects upon social, political, educational, and economic institutions; value and belief systems; and languages and artifacts” [3].

Culture, as a human product, is profoundly related to a place or region where natural environment has a great impact on people. Place describes the complex interplay of climatological, biological, geological, and topographical features that create the differences we see around us [4]. Place identity defines who people are and defends people from settings that threaten who they are or want to be [5]. The identity of a place can be seen as part of one’s self-identity derived from everyday experiences of places and the built environment. The sense of place identity is related to the meaning of that place to someone as “...place makes memories cohere in complex ways” [6].

Christian Norberg-Schulz, the architect, author, educator, and architectural theorist, relates cultural identity to its place through architecture as the manifestation of people and the ambient environment. In Norberg-Schulz’s writings runs, “...the unquestioned assumption that architecture has an identifiable ‘essence,’ the understanding of which is essential both to the discourse and practice of architecture” [7]. In his book “Genius Loci: Towards a Phenomenology of Architecture,” Norberg-Schulz states that “Human identity presupposes the identity of place.” The “essence” of architecture is defined accordingly: The basic act of architecture is therefore to understand the “vocation” of the place [8].

Architecture, vernacular in particular, is a product of people, place, and culture; it is one facets of identity. Symbolism of architecture can be related to the realization of identity personally and socially. This accretion has reached the level where “architecture as identity” became the equivalent to “architecture as space” and “architecture as a language” ([9], P. 137). Architecture, as the most obvious physical artifacts of any culture, has the most to draw from and respond to the uniqueness of place [4].

3. Vernacular architecture: localism and cross-cultural effects

Vernacular architecture can be defined as unpretentious, simple, indigenous, traditional structures made of local materials and following well-tried forms and types [10].

In his book “Architecture without Architects,” Bernard Rudofsky, the writer, architect, collector, educator, designer, and social historian commented, “There is much to learn from architecture before it became an expert’s art” [11]. To comprehend people's experience with ambient natural and built environments, we need to recognize the cultural meanings of architecture in the society day-to-day life practices and experiences [12].

For me, vernacular architecture is the built environment (city, architecture, and interior spaces) created based on the society needs. It is built in accordance with the natural environment (geography, topography, site, climate, local building materials, labor experience, and building techniques) fulfilling people's physical, economic, social, and cultural norms. Vernacular architecture is a sign of identity; it is the "mirror" of nations that reflects place, time, and culture. Architecture is built by people to people; it has developed through time and modified itself through trial and error to fulfill society's needs in harmony with the ambient environment.

Due to restrictions in transportation means, vernacular architecture depended on local materials and skills; this led to conserve resources and created uniqueness and identity to each region's architecture [13]. Each material embedded its physical and aesthetical characteristics that dictated the architectural technology that fits to this material.

Every society that created architecture has evolved its own forms that adhered to people as its language, clothing, customs, and traditional stories. Until the collapse of cultural frontiers in the twentieth century, there are distinctive local shapes and details in architecture as a natural product of materials, technology, environment, and people's culture [14].

Brick, for instance, as one of the most ancient and popular materials used in hot arid regions such as the Arab World, required certain technologies based on its shape, size, and durability. Most brick units' dimensions are set to fit human being's hand; this was convenient when building walls and piers while the thickness of these vertical elements varied based on its constructional location, height, and structural loads. To solve the roofing problem, builders had to create new forms in accordance with brick's physical characteristics; arch was the brilliant solution. From arch motion, shapes that are more convenient were created; when arch moves horizontally, the vault is created and when arch rotated around itself, we had the dome. These creative shapes worked very well with the ambient environment and climate esthetically and functionally.

By time, theses shapes (arch and dome in particular) gained symbolic meaning beyond functionality and esthetics; they became part of the "cultural memory." As Harry Francis Mallgrave, architect, scholar, and editor declared: "One's memory image of an architectural style is admittedly built on the stylistics images of one's culture and these images gradually obtain a progressive clarity or strengthening, which elicits greater delight as a particular style advances" [15].

Esthetically, brick unique beauty worked well with the long sunny days in hot arid regions. Creative decoration with brick such as ornaments, calligraphy, access and recess, niches, corbels, and muqarnas exposed the effective visual beauty. Sun and clear sky revealed esthetics of brick ornamentation through the contrast of shade and shadow. Even though, some of these forms were created for structural and functional purposed, the aesthetical and creative aspect was not neglected (**Figure 2**).

Regionalism in architecture reflects local features related to place, culture, climate, and technology in a certain era; the result a timeless architecture [16]. If imported materials and high technology are not used smartly, then architecture, as a place definer, will lose relation to its region and ambient environment [4].

Kavous Ardalan in viewing the relation between globalization and culture states: "... analysis of globalization and culture necessarily requires a fundamental understanding of the worldviews underlying the views expressed with respect to the nature and role of globalization and culture" [17].

Figure 2.
Aesthetical effect of shade and shadow in vernacular architecture (Source: Salman, Maha).

Arab countries have started to reexamine their own traditions in search of their "own" unique values and principles. This process had an impact on the production of contemporary architecture and eventually triggered an intense discussion about how "localism" should be created other than copying fragments from the past.

Most of modern cities lack identity due to imported global styles and techniques that do not cope with ambient environment and do not reflect the uniqueness of each city/country and its people and society. Jason McLennan notes "A simple look at most of our communities being built today shows that we have indeed lost our respect for place" [4].

Since the mid-twentieth century, architecture in the Arab World that can reflect local identity by reviewing the potential of the region has started; it became a new approach for regionalism in architecture. This was one trend to regain identity in architecture as many buildings have been achieved; however, most of these buildings did not capture the essence and dynamic context of local identity. Generally, regionalism in architecture was summarized with certain forms and architectural features or materials with a nostalgic view to heritage without deep exploration to the core of vernacular architecture and local identity.

4. Sustainability, vernacular architecture, and identity

In the past three decades, the world witnessed great awareness toward environmental dilemmas such as global warming, resources depletion, energy, air and water pollution, waste, population growth, and globalization. These issues come within the responsibility of arguments on sustainability indicating the necessity for sustainability integration in the ways we live, act, use of resources, and build. In meeting the human needs, sustainable architecture represents the interrelationship between natural, cultural, social, and economic resources to create optimum relationships between people and their environments [18].

The word "sustain" means "to hold up" or "to support from below," a society needs to be supported by its populations in the present and future. Most traditional societies took care of their community through the amalgamation of natural and physical environment with cultural, social, and mystical values. These are the societies where sustainability is evident and livable as daily practices.

When discussing sustainability, we need to go beyond definitions, looking for the essence of the meaning. Sustainability is not merely the integration of environmental, social, and economic issues and improving quality of life [19]. The idea of sustainability is to make sure that our decisions and actions today do not compromise the future generations to meet their own needs. Re-evaluating and analyzing vernacular architecture support the interrelation between policy makers and the society [20]. We have to make sure that our actions go along with our environmental systems not in opposition to them. Sustainable design implies responsibility and it implies a far-reaching respect for natural systems and resources, respect for people and cycle of life [4].

Sustainability is by nature "architecture." If we are to achieve it, we must concretely engage the design issues associated with ambient environment resources. We must address, more fully, the underlying influence(s) of political, economic, and social issues comprising the cultural and spiritual background in which the desire to achieve sustainability in itself reflects such a significant value shift [21]. Sustainable design is a philosophical approach to design that seeks to maximize the quality of the built environment [4].

Vernacular architecture is the result of constrains on resource availability. Historically and without modern means, extraordinary enterprise produced architecture often of the most distinctive character and identity with only limited means available [22]. Vernacular architecture was built on inherently sustainable principles such as resource limitations imposed by economic or natural factors, yet succeeded in offering rational solutions to harsh climates and human need [23]. Elements of sustainable design are integral to every established form of vernacular architecture that are tried and proven solutions and have evolved over long periods of trial and error using local materials and technology emerging from ambient natural and cultural environment.

Vernacular architecture, in the Arab World, is a model for sustainability; it embodies different cultural values, which may be applied in the conceptual design of buildings today. These values are about the way of construction, which is responsive to the climate and suits the style of living, traditions, customs, social values, and culture of people; it is of a great sustainability potential (**Figure 3**).

Sustainability in architecture is an essential part to determine the identity of the architectural regionalism. Hidayatun, Prijotomo, and Rachmawati assert that: "Identity has a permanent nature, permanence is supported by both natural environment and cultural, such as the natural topography, material and cultural includes views and habits. All of this is contained in the sustainable criteria. So the nature of sustainability is an important part in recognizing the identity" [16].

Figure 3.
Accommodation to the ambient environment in vernacular architecture in the Arab World: solid walls, small openings, light finishing, courtyards, and wind catchers (Source: Salman, Maha).

Since our predecessors built sustainability, they did not need to enforce methods to reveal their society or place identity. Identity was evident by being adhered to natural and social environment while society evolution through time. They built naturally in accordance with ambient environment and available resources and building materials; architecture embraced identity as it reflected its place, people, and culture.

5. Vernacular architecture revitalization: interpretation of identity

Tradition is portion of a society culture that transferred from one generation to another. Apparently, a society's identity is integrated within its traditional heritage resulting on placing great value on heritage. "Tradition is not necessarily old-fashioned and is not synonymous with stagnation. Modernity does not necessarily mean liveliness, and change is not always for the better. On the other hand, there are situations that call for innovation" [14].

In the Arab World, tradition is not a sole layer of heritage or culture; it is multi-layered complex. Recognizing tradition, as a product of society, requires a creative transformation of its cumulative values to next generations incorporating contemporary values that did not exist before [24].

5.1 Vernacular architecture, people, and society

One of the most fruitful developments in architectural thoughts to emerge in the last three decades is the changed perception of tradition and innovation and the growing realization of their interdependence. Making connections between existing but previously unrelated concepts is now understood to be as much a part of the creative process as having a free and lively imagination. Nothing comes from nothing, it could be said, and tradition is both a potential launch pad for new ideas as well as a potential hindrance. The present task therefore is to better understand the reciprocal relations between the two, the way one complements, feeds into, stimulates, or hinders the other [9].

The cultural and social core of vernacular architecture need to be seen within the perspective of routine practices of people and community to grasp the essence of society's experience with the built and natural environment. Many studies related vernacular architecture to natural and cultural environment of the location and region. Joseph Kennedy defines vernacular architecture: "an architecture style that develops from the particular climate and social conditions of a place" [25].

In his book "Encyclopedia of Vernacular Architecture," Paul Oliver, an architectural historian and writer, accentuated the multidimensional relation between society and vernacular architecture: "It is particular characteristics of vernacular architecture that each tradition is intimately related to social and economic imperatives; it has developed to meet specific needs within each cultural milieu" [26].

Predicatively, vernacular architecture offers the greatest potential for the development of a viable contemporary regionalism of consistent of high quality, capable of providing for many building types, both old and new. The potential diverse from the sheer richness of the heritage diversified over centuries of continuous development [9]. Modern life styles and advanced technology necessitate innovation; however, innovation and creativity must be deeply thought-out response to the change in circumstances and not just indulged to its own sake [14].

5.2 Vernacular architecture: search for identity

During architectural modernization in the Arab World starting 1950s, many traditional buildings and heritage symbols were demolished for they were symbols of retardation and poverty. In paving the path for modernization in Arab cities, architectural heritage was abandoned and expertise of its construction lost. Many historic buildings suffered degradation and decay due to neglect and misuse by informal development and occupation by less fortunate people.

In spite of the several demands by professionals, public, and suggested policies in 1970s and 1980s on the conservation of the traditional urban fabric of the Arab city, the traditional fabric was demolished, partially or completely in favor of Western building style as a trend to modernity. This was one of the actions leading to loss of modern architecture identity in the Arab World. The new urban planning strategies were set to accommodate the modern life vibrant style where grid-iron planning, big streets, cookie-cutter plots, new construction materials, and high-rise buildings replaced traditional fabric neglecting cities' architectural heritage and society's cultural values.

Figure 4.
Modern buildings using traditional forms and architectural features (Source: Salman, Maha).

It is a fact that when traditional city core demolished drastically, this affects people's social behavior and may affect their feeling toward their city and life style [27]. Rapid culture change enables one to identify core versus peripheral elements, those elements changing rapidly versus those remaining relatively constant.

In an attempt to reveal identity in modern architecture in the Arab World, some architects, local and international, began to use elements, forms, and motives in particular, from vernacular architecture in the design of new buildings as a resemblance of the (glory) of the past. Some modern architectural trends in the Arab World, in efforts to "reveal" identity, have defined the local architectural identity by demonstrating distinctive characteristics of vernacular building elements to be symbols of local identity [28] without deep analysis of the real essence of how these elements originated and by what means they reflected people, place, and environment.

This approach became a (facial) mask; it was mainly evident on facades, openings, and architectural features such as arches, ornaments, and domes rather than the design concept or interior spaces (**Figure 4**). New construction materials and technology such as concrete, steel, big glass opening, and air-conditioning became essential in most new buildings with total abandonment to local materials and ambient environment.

There were a few attempts to be inspired by the core design elements of vernacular architecture such as environmental solutions. However, there were several examples for regional architecture, designed by western and Arab architects trying to create an alternative to the prevailing international models [24] such as the work of Mohammed Makkiya, Hassan Fathy, Rasem Badran, Jafar Tukan, Rifat Chadirji, and Elie Azagury. This awareness of the importance of vernacular architecture and tradition to enhance the built environment was not merely a nostalgic approach to the past; it was about endurance of local identity.

5.3 Vernacular architecture: sustainability and identity

Vernacular architecture is of great potentiality to modern "...as the late twentieth-century people, though, limitations of traditional cultures in helping us to know what to do are evident" [29]. Nevertheless, vernacular architecture features offer dramatic metaphors for regional forms of shelter, as well as rational responses to the harsh climate giving modernism a subtle but telling shove in direction of regionalism [9].

The multiplicity of sustainability trends has a great impact on the architectural identity and image of the Arab city. Abdul Salam and Rihan defined three trends: the modern technology, the neotraditional, and the contemporary interpretation trend investigating reasons of their emergence, main features, and impacts on the architectural identity in Arab World [30].

Revitalization of architectural heritage and conservation cultural values in an environmentally sensitive manner requires detailed planning. It needs knowledge of materials and their interactions; knowledge of construction, craft techniques, skilled technicians, and available resources; and an ongoing commitment to the region identity through a sustainable approach. Heritage revitalization within sustainable approach needs an in-depth interpretation of core values of this heritage that goes beyond the typical "imitation" and visual metaphors of traditional forms and architectural features and ornamentations.

Vernacular architecture has the essence of sustainability to be the link to more environment-respected buildings. To ensure society continual survival, we need to adapt sustainable solutions for growth through respecting the environmental balance of nature. The reevaluation of vernacular architecture can offer an indefinite source to develop worthwhile ecological solutions for the built environment responding to society needs [31].

It is not just nostalgia that draws people to vernacular architecture. Much of what is valued in this architecture is its sustainability and response to the climate, natural setting, and locally available building materials. Their usefulness as model for new buildings only adds to their value.

6. Sustainability and vernacular architecture: redefining identity

In traditional societies and for centuries, people have lived in harmony with nature; they grew their food from surrounding region and developed their lifestyle in accordance with the available resources. They have construct buildings using the local building materials available in the surrounding environment using their hands and developing building techniques affiliated with the physical characteristics of these materials. Traditional societies realized that their survival required them to sustain balance with lifecycle around them. In other words, traditional societies were the real pioneers of sustainable development over time in the perspective of natural and built environment.

When defining architecture and the built environment, we have to think beyond places shaped by products and buildings; we need to consider how people modify the environment to accommodate to their needs, life, customs, and culture. McClure and Bartuska noted that "The Built Environment is everything humanly made, arranged or maintained to fulfill human purposes, needs, wants and values to mediate the overall environment with results that affects the environmental context" [32]. This can be related to how sustainability is integrated with architecture.

The architectural style, design, and construction materials of new buildings should reflect the cultural heritage of the locality or region and they should be environmentally and culturally sensitive and sustainable over the long term (**Figure 5**).

The architecture of the region, created at specific geographic locations at certain points in time by different individuals, reflects spirit of identity. There is an interactive relation between sustainability and identity. Sustainability, in its comprehensive essence, has integrated within physical and incorporeal heritage of any society. Social and environmental sustainability has incorporated with local identity where people formed physical production (buildings, artifacts, furniture...) and cultural values (identity, traditions, social values...) in response to the ambient environment. Architectural identity, and its interrelation with the context such as location, climate, environment, local material, encloses the meaning of sustainability: "... sustainability is important part of the identity in architecture; identity is the main core in the dimension of regionalism architecture" [16].

Figure 5.
Inspiration from vernacular architecture within urban context (Source: Salman, Maha).

Figure 6.
Inspiration from environmental solutions in vernacular architecture within modern urban context-Masdar City-Abu Dhabi-UAE (Source: Salman, Maha).

Sustainable architectural is built on the idea of regionalism; however, it does not ask us to return to the old ways of living as pure nostalgia; it inspires us for the responsible, long-term use of technology and design [4]. Interpretation of architectural styles through features such as: design concept, inward looking plan, thick brick or stone walls, small well-designed windows, door transom, high ceilings, and elevation treatments can provide valuable lessons in sustainability. How these low-tech features functioned during times when energy consumption was limited provides examples of principles applicable to today's efforts to conserve local identity as a response to environmental issues. The traditional response to climate, setting, and materials provides opportunities for presenting positive lessons in environmentally sound design.

Preserving the identity in the Arab World architecture necessitates a deep understanding of the local natural systems and thorough perception of cultural values of the society that have developed and proven their validity over time. Architecture needs to be erected to meet the desires and needs of its people; the built environment is shaped with unblemished consciousness of the multilayered relations between nature, culture, social values, economy, and available resources of the region and its distinctiveness.

Sustainable design philosophy honors the differences between places; it rejects the concept that buildings should look and being built the same in any region. The making of cultural identity in contemporary architecture reveals the mechanisms of constructing the "regionalism" and the "identity" through architecture in an international, global context. Architecture should respond to place in fundamental ways [4].

Advocates of sustainability acknowledge that traditional cultures and ideologies embody ideas and principles on which sustainable living should be based [33]. The goal of sustainable building design is to create optimum relationships between people and their environments. Planners, architects, designers, developers, and operators have an opportunity and a responsibility to protect the identity of a place, its people, and its spirit (**Figure 6**).

It is not just the use of appropriate materials or local building techniques, or the reevaluation of traditional features need to be preserved; it is the identity of people has to be sustained in order to overcome the exploitation and the cultural neocolonialism which plays such a great role in contemporary Arab World architecture.

7. Conclusions

This chapter intended to discuss identity in architecture through investigating several concepts such as culture, vernacular architecture, and sustainability as different facets of identity. Several definitions have been discussed for the major concepts of this chapter.

Table 1 illustrates a comparison between some definitions as well as highlighted author's interpretation.

Through discussion and analysis, the chapter tried to find analogies between identity, as an indicator of a certain society, and sustainability, as a lifestyle developed by people in that society to live in harmony with the surrounding nature and available resources sustaining balance with lifecycle around them. Architecture, vernacular in particular, as a product of people is a vibrant manifestation of how identity and sustainability are related spontaneously in accordance with nature and culture.

Vernacular architecture is a demonstration of identity and sustainability; it is the "mirror" of nations that reflects place, time, and culture. Architecture that was built by people to people; it has developed sustainably through time and modified itself through trial and error to fulfill society's needs in harmony with the ambient environment.

The search for establishing a new regional identity means being independence from the imported values and ideologies without losing cultural interrelation that serve the human civilization. Regaining an Arab identity is fundamental to the formation of a new Arab culture, not only in the field of architecture but throughout all aspects to feature a trace in the drastic changeable world.

Identity
Identity is the fact of being who or what a person or thing is; the characteristics determining who or what a person or thing is [1].
Identity establishes or indicates who or what (someone or something) is, who a person is, or the qualities of a person or group that make them different from others [2].
Human identity presupposes the identity of place. The "essence" of architecture is defined accordingly: The basic act of architecture is therefore to understand the "vocation" of the place [8].
Place identity defines who people are and defends people from settings that threaten who they are or want to be [5].
The identity of a place is part of a one's self-identity derived from everyday experiences of places and the built environment [6].
Identity means being unique and distinguished from others, this can be applied to a thing, person, and group of people, society, country, or even nation.
Vernacular architecture
Vernacular architecture is unpretentious, simple, indigenous, traditional structures made of local materials, and following well-tried forms and types [10].
Vernacular architecture is an architecture style that develops from the particular climate and social conditions of a place [25].
In vernacular architecture, each tradition is intimately related to social and economic imperatives; it has developed to meet specific needs within each cultural milieu [26].
Vernacular architecture features offer dramatic metaphors for regional forms of shelter, as well as rational responses to the harsh climate [9].
Vernacular architecture is the built environment (city, architecture, and interior spaces) created based on the society needs. It is built in accordance with the natural environment (geography, topography, site, climate, local building materials, labor experience, and building techniques) fulfilling people's physical, economic, social, and cultural norms.

Table 1.
Summary of identity and vernacular architecture definitions.

Conceptually, sustainability is not all about energy conservation, renewable resources, or building materials. Sustainability is a life style, a way of living, and society cultural identity. Human being everyday activities are the essence of sustainability, the way they live, act, work, produce, plant, and build. Sustainability is not a term to be understood; it is a life to be practiced.

Regaining identity requires deep investigation of vernacular architecture to contrive potential of sustainability in hardpan. Sustainability, as a style of living, is one of the routes in our search for identity in this global context, which is indeed a search for survival, not only for the Arab World but also for the entire world.

Conflict of interest

The author declares no conflict of interest.

Declaration

All photos were taken by Dr. Maha Salman.

Author details

Maha Salman
Theory and Sustainable Architecture, Yorkville University, Toronto, Canada

*Address all correspondence to: mahazubaidi@yahoo.com

References

[1] English Dictionary, Thesaurus, & Grammar Help|Oxford Dictionaries. Oxford Dictionaries|English. Available from: https://en.oxforddictionaries.com/definition/us/identity/identity [Accessed: July 21, 2018]

[2] Identity Meaning in the Cambridge English Dictionary. Gender Pay Gap Definition in the Cambridge English Dictionary. Available from: https://dictionary.cambridge.org/dictionary/english/identity [Accessed: July 21, 2018]

[3] Jani V. Diversity in Design: Perspectives from the Non-Western World. New York: Fairchild Books; 2011

[4] McLennan JF. The Philosophy of Sustainable Design. Kansas City, MO: EcoTone; 2006. P.6, 10, 52, 53

[5] Kopec DA. Environmental Psychology for Design. New York, NY: Fairchild Books, an Imprint of Bloomsbury Publishing; 2018

[6] Gieseking Jen Jack, Mangold William, Cindi Katz, Setha Low, Susan Saegert. "Section 3: Place and Identity". The People, Place, and Space Reader. Available from: https://peopleplacespace.org/toc/section-3/ [Accessed: July 04, 2018]

[7] Abel C. Architecture as Identity, I: The essence of architecture. In: Herzfeld M, Lenhart MD. Semiotics 1980. New York: Plenum Press; 1982

[8] Norberg-Schulz C. Genius Loci: Towards a Phenomenology of Architecture. New York: Rizzoli; 1996

[9] Abel C. Architecture and Identity: Responses to Cultural and Technological Change. London. New York: Routledge; 2000

[10] Curl JS. A Dictionary of Architecture and Landscape Architecture. Oxford: Oxford University Press; 2006

[11] Rudofsky B. Architecture without Architects: A Short Introduction to Non-Pedigreed Architecture. Albuquerque: University of New Mexico Press; 2003

[12] Salman AZ, Maha. The sustainability potential of traditional architecture in the Arab World. Doctoral [thesis]. Huddersfield: University of Huddersfield; 2007

[13] Winchip SM. Sustainable Design for Interior Environments. New York: Fairchild Books; 2011

[14] Fathy H. Architecture for the Poor. Chicago: University of Chicago Press; 1976

[15] Mallgrave HF. Architecture and Embodiment: The Implications of the New Sciences and Humanities for Design. London: Routledge; 2013

[16] Hidayatun MI, Prijotomo J, Rachmawati M. Sustainability is important part of the identity in the dimension of regionalism architecture. Applied Mechanics and Materials. 2015;**747**:145-148. DOI: 10.4028/www.scientific.net/amm.747.145

[17] Ardalan K. Globalization and culture: Four paradigmatic views. International Journal of Social Economics. 2009;**36**(5):513-534. DOI: 10.1108/03068290910954013

[18] Salman M. Sustainable house... Futuristic approach for sustainable architecture. In: Proceedings of Housing Symposium II. Riyadh: Al-Riyadh Development Authority. PP; 2004. pp. 422-435

[19] Senosiain J. Bio-Architecture. Oxford: The Architectural Press; 2003

[20] Hinrichs CL. The architecture of the courtyard house as generated by passive solar energy & other factors. MPhil [Thesis]. Faculty of Arts, School of Architecture. Manchester: The University of Manchester; 1987

[21] Koester RJ. Sustainability is an architecture. Architronic. 1995

[22] Thomas D. Architecture and the Urban Environment: A Vision for the New Age. Oxford: Architectural Press; 2002

[23] Salman M, Easterbrook S, Sabie S, Abate J. Sustainable and smart: Rethinking what a smart home is. In: Proceedings of ICT for Sustainability; 31 August-2 September 2016. Amsterdam: ICT4S; 2016. DOI: 10.2991/ict4s-16.2016.22

[24] Kultermann U. Contemporary Architecture in the Arab States: Renaissance of a Region. New York: McGraw-Hill; 1999

[25] Kennedy JF. Building without Borders: Sustainable Construction for the Global Village. Gabriola, B.C.: New Society Publishers; 2004

[26] Oliver P. Encyclopedia of Vernacular Architecture of the World. Cambridge: Cambridge Univ. Press; 1998

[27] Bokhari AY. On the identity of the Arab–Islamic city: Past & present. In: Serageldin I, El-Sadek S, Herbert RR, editors. The Arab City: Its Character and Islamic Cultural Heritage. Arlington, VA: I. Serageldin; 1982

[28] Sani RM, Shotorbani PM. Symbolic Use of Wind-Catchers in Iran. Gateshead: Open House International 2013;**38**(2):76-87

[29] Farmer J, Richardson K, Bradley JF. Green Shift: Towards a Green Sensibility in Architecture. Oxford: Butterworth-Heinemann; 1996. p. 12

[30] Abdelsalam T, Rihan GM. The impact of sustainability trends on housing design identity of Arab cities. HBRC Journal. 2013;**9**(2):159-172. DOI: 10.1016/j.hbrcj.2013.03.002

[31] Klinker S. Shelter and sustainable development. In: Kennedy JF, editor. Building without Borders: Sustainable Construction for the Global Village. Gabriola, B.C.: New Society Publishers; 2004

[32] McClure WR, Tom J. Bartuska. The Built Environment: A Collaborative Inquiry into Design and Planning. Somerset: Wiley; 2011

[33] Mortada H. Traditional Islamic Principles of Built Environment. London: Routledge; 2011

Chapter 5

Repackaging "Traditional" Architecture of the African Village in Zimbabwe

Umali Saidi

Abstract

Repackaging is being done through construction of buildings with connections to traditional architecture as well as respective traditional naming of such symbolizing a return to African values. Naming of spaces and buildings is closely tied to the architecture of African heritage. This chapter explores the architecture of selected outdoor spaces in Zimbabwe. It is argued that the new places symbolize, represent African culture in terms of homage, food, hospitality, and cultural security. Furthermore, it is argued the spaces offer an eyrie of cultural values as housed in the architecture as well as names of the spaces thus communicating African sensibilities, aura of the traditional village at the same time sustainably conserving the architecture authored as by African heritage. One of the conclusions made is that the urban landscape can store and communicate African heritage; interestingly, requisite architecture can be used in this endeavor. Data were gathered qualitatively, and participant observations, interviews and extensive desktop research were involved.

Keywords: semiotics, village, hut, names, architecture, heritage

1. Introduction

For years, Africa has rolled out efforts to assert its position within a globalized "village." In some cases, the continent has been presented as torn apart between having either to appropriate new systems and risk annihilation by western systems (and culture) or to recreate its systems enshrined in its culture (or indigenous knowledge systems) in order to regain its rightful place in a globalized world. Globalization, as Yankuzo [1] argues, is a critical tool for cultural homogeneity designed to end cultural diversity in the world. This means, diverse cultural values are replaced by global cultural values, and the varieties of social and cultural developments have been fuelling globalization over the years. Furthermore, Yankuzo [1] believes globalization as a phenomenon has "a multi-dimensional process, whereby cultural, economic and political relations increasingly take a global basis" (p. 2). Against this view, what it means is that Africa continues to negotiate for space in this globalized space. Africa's traditions, practices, and discourses reflect that the continent has not been annihilated by globalization. Instead, Africa continues to mark its unique space in varying ways.

In Africa, the "rural landscape" has been characterized in the discourses as the source of African sensibilities. The African village is a granary of that which is African [2], with the urban landscape providing contrasting images tied to westernization, urbanization and all forces of modernity. African indigenous knowledge systems have extensively shown how the rural landscape has been the sanctuary of African culture, traditions and values. Urbanization, in contrast, has been projected as a threat to these values. One reason behind the continued smoking of the *ncelwa*[1] and practices attached to the *buntibe*[2] by the BaTonga people of Zimbabwe has been that the BaTonga were "neglected" for years [3]. Ironically, this "neglect" may be read as part of the many reasons why, a great deal of their culture remains "uncontaminated" or less contaminated by "modernity."

Even African literature, reflects aspects of cultural erosion. For instance, Lucifer in *Waiting for the Rain* [4] proclaims that his birth in Africa was against his will and ultimately becoming an African was to him a biological error. Lakunle in *The Lion and the Jewel* [5] detests, like Lucifer, African values as barbaric, primitive and excommunicated practices. The degree of cultural erosion and westernized is so vivid that writers call for efforts in "returning" to the source of cultural values and practices especially because contemporary African urban dwellers have rural backgrounds [6] or have the village as their place(s) of origin. Most Africans moved to the urban landscape for economic reasons.

This chapter, therefore, architecturally explores selected outdoor spaces in Zimbabwe. The chapter argues that the new places symbolize and represent African culture in terms of homage, food, hospitality and cultural security. Furthermore, it is argued that while this triggers nostalgia to Africans, the spaces offer an eyrie of cultural values as housed in the architecture as well as names of the spaces in order to communicate African sensibilities about the African aura of the traditional village at the same time sustainably conserving the architecture authored by African heritage. One of the conclusions made is that the urban landscape can store and communicate African heritage, interestingly, requisite architecture can be used in this endeavor.

The chapter also argues that the naming and architectural systems also reflect modernization as well as a sense of economization of the African village since those behind the naming and constructions are taking advantage of the economic opportunities presented by the twenty-first century land reform program. It is further shown that African culture has thus survived or is given a lifeline by being repackaged, modernized and appropriated in the new terrain; a terrain where modernity forces and processes have in some way sort over the years to delete ultimately, African sensibilities.

Data for the study were gathered qualitatively. Interviews with owners (and consumers) of selected lodges and restaurants were conducted during visits undertaken to such spaces by the researcher. Participant observations were also done as the researcher participated in various activities organized at such spaces such as weddings, workshops, parties, and various other social activities. Furthermore, extensive desktop research was conducted to enrich the study. In order to lay the foundation for my discussion and reporting, first, an aside on the twenty-first century Land Reform Program in Zimbabwe is done to put the discussion into context.

---

1 BaTonga women smoking pipe

2 BaTonga drum(s) and cultural dance

2. Twenty-first century land reform in Zimbabwe

Documentation of the twenty-first century land reform, dubbed the Fast Track Land Reform Program (FTLRP) in Zimbabwe, is quite visible, and there is a considerable corpus of studies on the land issue in the country. Some scholars have called land issues in the country, *Zimbabwe's Unfinished Business* [7]. However, most studies on FTLRP generally agree that the program was a result of the country's thirst to settle the land question in the country. Others have provided varying perspectives on the land issues with the likes of Vambe (ed.) [8] deciding to deal with political dynamics involved in the program, which subsequently affected economic and social lives of the Zimbabwean people, given the heavy political undertones characterizing it.

In her report, Tibaijuka [9] reveals the many sidedness as well as many faces created by the land reform in the country. For instance, there are narratives detailing the dynamics of Operation *Murambatsvina* (Operation Restore Order), dubbed "Operation Tsunami" because of its speed and ferocity. Fontein [10] recounts realities during the operation when ordinary people in the cities destroyed their own residential structures considered "Illegal." Studying the aftermath of the FTLRP, Potts [11] observes the economic and humanitarian crisis and dislocations that followed the land reform program. More broadly, however, Sadomba [12] reflects the historical underpinnings of land reform from 1980 and the agitation as well as demand for land by war veterans. Moyo and Chambati [13] conclude that the FTLRP represents the only instance of "radical redistributive land reforms since the end of the Cold War," reversing,

> *... the racially-skewed agrarian structure and discriminatory land tenures inherited from colonial rule, whereby over 6000 large-scale white farmers and a few foreign and nationally-owned agroindustrial estates controlled most of the prime land, water resources, and bio-reserves, while relegating the majority of the population to marginal lands and cheap-labor services. ([13], p. 1)*

The twenty-first century land reform program in Zimbabwe also attracted the attention of the international community and imagination. For the locals, it rekindled memories of the past mostly tied to colonialism and its land policies. Mutondi [14] notes that for the Zimbabwean people, it was also about economic empowerment read as "the final embodiment of empowerment following Zimbabwe's independence in 1980" (p. 1). Political interpretations of the land reform program were two sided. Drivers of the program claimed the program was overdue. The then President, Robert Mugabe, declared in 2005 that

> *without doubt, our heroes are happy that a crucial part of this new phase of our struggle has been completed. The land has been freed and today all our heroes lie on the soil that is declaration. Their spirits are unbound, free to roam the land they left shackled, thanks again to the Third Chimurenga. ([14], p. 2)*

Defenders, mostly beneficiaries of the land reform (mostly politicians), projected the program as successful. The defense was centered on the recentralized "decision-making relating to land matters, ostensibly because [the government] perceived itself as fighting against external forces bent on reversing the gains of the program" ([14], p. 4). Thus, political statements and action were pitched very high in order to hide certain local tendencies (some of the action relating to the land takeovers was permeated by violence), which made it difficult to discern what was happening on the ground ([14], p. 4).

Chief among the defenders of the program was the government, its institutions, agents, and officials. But those opposed to the government's model placed at the center "human rights violations" along racial, sometimes tribal lines. The white Zimbabweans were characterized as victims, and black supporters of this were also seen as victims too. The Mugabe government characterized black people who opposed its stance and model of the program as lacking totems, connections to the Zimbabwean culture and identity; stooges meant and being used to reverse the gains of the armed liberation struggle that brought independence in 1980.

Perhaps, the politically charged periods of 2000–2008 rendered the program more political than economic or sociocultural. As such, studies done during the same period tended to take the same lines with conclusions being submitted as the program being unsustainable rather than a socially and economically sustainable developmental drive for the people of Zimbabwe. Almost two decades after the FTLRP, research and discourses now center on measuring the success of the land reform based on arguments around poverty reduction and economic efficiency with political considerations gradually wearing away.

Regardless of the cost in humanitarian sense, economic, political, and social complexities as well as dynamics, the program created economic opportunities for the locals around land ownership. Not only did the land reform address agrarian related land ownership and utilization, but residential as well as commercial land was also made available for expansion of existing settlements as well as for creation of new settlements and economies. It is this reading of the program, which opened up economic opportunities in terms of the availability of commercial land, which this chapter heavily sides with. The new spaces have allowed repackaging of "tradition," specifically repackaging of the traditional African village architecture and subsequent naming of the created spaces since most owners became locals.

3. Semiotics, architecture, and names

Buildings and their subsequent names are semiotic resources. Semiotics is used to view buildings as objects as well as respective place names as resources in the construction as well as naming patterns and systems of the landscape itself. Not only does this empower us to interpret meanings carried by lodges and restaurants, but also this alludes to the language of the landscape as it has come to communicate a variety of messages.

When linguistic sign systems are used either as names, strings of sentences, or lexical items, the sign systems create a signifying system. Ikegami ([15], p. 1) calls such a system a "meaning-generating" system "observable in the cultural spheres they are dealing with" ([15], p. 1). The science (sometimes approach) dealing with signs, their systems, and general function in meaning making as well as interpretation is known as *semiotics*.

Broadly speaking, semiotics is a phenomenon in which something is used to stand in for something else. De Saussure [16] and Pierce [17] used "semiology" and "semiosis," respectively, which, in modern semiotics, are basically understood as semiotic processes. With reference to "semiosis," [15, 18] human beings operate and survive by evaluating as well as assigning significances and values to things, phenomena "around him *in relation to himself*"[3] ([15], p. 2). This means, human existence is determined by use of sign systems where the world becomes a network of signs represented, or (mis)represented, as a network of sign systems.

[3] Emphasis original.

Semiotic knowledge teaches us that anything can function as a sign since signs do not have predetermined properties. This also means that buildings and their names do not have predetermined properties as such function as semiotic signs. However, the observable properties of signs are in fact defined in terms of their relation to what they represent. Johansen and Larsen ([19], p. 26) say signs are an "inconspicuous part of our everyday existence" revealing a triadic relation among sign, object, and interpretant. *Icons*, *indexes,* and *symbols* are given as signs in a bid to explain the conditions of representation (or otherwise) and meanings generated when signs function [17]. More broadly, therefore, we can borrow Braga and Mundy [20] who, in attempting a position on whether semiotics is a science or a method, say:

> *...semiotics can...be thought of as an applied science, in that it serves as an instrument for the study of empirical objects, or rather for the description and analysis of concrete products of language, whether they be a poem, a theorem, a piece of music, a utilitarian object, a rite, a political speech, a play, an art object, a movie, a TV program, etc. In short: each and every linguistic manifestation. (p. 115)*

From the above, we note that aspects from which meaning making especially through language and its use is fertile ground on which semiotics can be and has been used. Riffaterre (in Champagne ([21], p. 229)) used semiotics to analyze French poetry for the simple reason that "poems are ... assumed to be closed units of meaning ..." Hale [22] speaks of Umberto Eco as having used semiotics in the writing of his novels notably *The Name of the Rose* [23], *The Island of the day before* [24], and *The Mysterious Flame of Queen Loana* [25]. Cunningham [26] submitted the relevance of semiotics in educational psychology and, hence, speaks of educational semiotics [27]. Dunbar-Hall [28] used semiotics as an analytical method to study popular music. The list, therefore, is endless and cannot be exhausted here, hence, by extension, we can revoke semiotics in analyzing onomastic data.

4. Buildings and their names as semiotic resources

Architectural objects such as buildings and their names within a landscape form a language. They are units of a landscape language. Their creation and use create patterns and communicative sign systems. Furthermore, buildings and space place names derive their energy from one who designs, constructs, and names them. This means there are even power dynamics as well as circumstances or events that may be involved in the designing, constructing, as well as naming of a thing or a place. The colonized participated too in various ways in naming as well as erecting buildings from their worldview, thus creating places and spaces in response to new realities they now experienced.

Places, spaces, and objects (buildings), therein, follow the triad and a variety of signs [16, 17]. We find them either as *icons*, *symbols,* or *indexes* depending with the ultimate meaning intended. In other words, using or handling buildings, spaces, and their names as semiotic resources means *meaning* is not fixed all the time; hence, the ability of these spaces at one point becomes *icons* on the other *indices* and yet on the other being *symbols,* but all the time confined within the sign-signifier-signified conception. From this background, therefore, lodges and restaurants in Zimbabwe at the backdrop of the FTLRP assume a similar character.

The practices and systems observable in contemporary Zimbabwe mark out the definition and "ownership" of spaces created by the land reform program. The program empowered the locals to design, construct, and "name" these new spaces. Remember, to name is to control; thus, the locals who partook in setting up the said lodges and restaurants also named new spaces as a reflection to the newly found economic opportunities. They have done so consciously aiming at creating indigenous spaces in order to evoke as well as communicate African sensibilities and their heritage.

5. Selected spaces and their description

In this section, I report as well as describe the architecture of some of the most prominent lodges and restaurants. While owners of these places are locals, the common names I observed as well as the outlook of their architecture had some resemblances distinguished mostly by locality and ownership. Both male and female locals have ownership of the spaces, and in some cases, this appeared to some extent to determine the naming patterns as shall be discussed below. The physical space location of the lodge or restaurant was noted to be the most distinguishing factor with ownership varying significantly. The following are some of the place names.

5.1 The village lodge

"The village lodge" was observed to be a common space for newly established lodges dotted in considerable outskirts of main towns and cities in Zimbabwe, mostly in Midlands and Mashonaland provinces. In terms of naming, what is key in this place name is the prefixing of the main economic activity "the lodge" with the lexical item "village." These spaces apart from offering overnight accommodation for travelers, they also have restaurants and conference rooms, hence, not only favoring for overnight accommodation, but also offering venues for the conduction of various other social activities such as wedding ceremonies, business workshops, and conferences. Owners did not hide their niche from the naming patterns as expressed grave desire in making a mark in the hospitality industry, and thus, the name "village lodge" was in itself suggestive of an African-centered ambient for the conduction of the above activities.

5.2 Chicken hut

"Chicken hut" in contrast to "the village lodge" was observed as primarily a restaurant or eating space where dominant meals served are poultry-centered one. Beyond offering restaurant facilities, hiring out of basic catering services are also offered for weddings and other gatherings, which may be staged far removed from their physical locality. Nonalcoholic beverages are also served with the said meals. The irony, however, lies in which the poultry-centered meals served are mostly based on foreign menus with Italian, French, and Greek as well as American being much popular. Physical location of the spaces was generally observed to be within the central business districts of the towns or cities.

5.3 *Pamuzinda* (The King's Kraal, King's Homestead)

"*Pamuzinda*" (The King's Kraal, King's Homestead) speaks volumes in terms of connotations to royalty. Like "chicken hut," these are mostly restaurant, but they have an extra niche to them which is the bar. Meals are mixed, with African traditional meals making the bulk of the dishes served, however, at the backdrop of a bar. One example is *Pamuzinda* at the outskirts of Harare, which is more of a bar and the space provides entertainment where even prominent Zimbabwean musicians are hired to perform for revelers

DOI: http://dx.doi.org/10.5772/intechopen.81450

KwaMaiguru/KwaGulez (my brother's wife's place),
PaMaGumbo, (a place of one of the *Gumbo*/Leg totem),
PaNdari (the Shabeen),
PaMusha (at the homestead).

KwaMaiguru/KwaGulez (my brother's wife's place) like *Pamuzinda* while offering revelers with food and drink (alcoholic mostly), the most characteristic feature of these is the open spaceness. The same also goes for *PaMaGumbo* (a place of one of the *Gumbo*/Leg totem), *PaNdari* (the Shabeen), and *PaMusha* (at the homestead). The central aspect, however, is the food, where traditional meals and menus are used. These are read as "eating open spaces" the common feature of the meal being "braai," served with *Sadza* (thick porridge). Alcoholic beverages are also served, mostly beer. Revelers participate in self-entertainment especially in as far as music provision is concerned as much of it is played from their vehicles.

6. The African village "reconstructed": architecture and naming

The "village" concept reveals African cultural values, which include and designed to reveal a sense of community life, a sense of good human relations, a sense of the sacredness of life, a sense of hospitality, a sense of time, a sense of language, and proverbs to mention a few. The vernacular architecture and naming of the above places against the "village concept," hence, signify and symbolize much of this African village concept. To appreciate the African values attached to the "village," we note that in terms of the "sense of community life" offered by the African village; traditionally, Africans believe that the community is the custodian of the individual. In the material reference to the phrase, the individual must thus return or go to the "community center" or village square, which is a social, political, judicial, and religious center.

Of the view that, "today, many urban Africans coming from a village and argue that it is in the village where traditional African culture can still be found" ([6], p. 51), it follows the locals in Zimbabwe after the FTLRP by ridding on the village concept are in fact aiming at establishing as well as producing locality in order to redefine themselves as "actors who properly belong to a situated community of kin, neighbors, friends, and enemies" ([29], p. 179). Lucifer, in *Waiting for the Rain* [4], returns to the "village" before undertaking his journey abroad. Much of the class of western and African cultures as well as reflections of the dilemmas that Lucifer faces are revealed by his being back "home," the village, which in this case is the "community center."

Furthermore, in terms of the "sense of good human relations" attached to the village concept, we learn that relationship between individuals reveal their worth as human beings, hence, the art of dialog, discussion of problems, and searching for solutions through dialog. Within the spaces presented above such as *KwaMaiguru/KwaGulez* (my brother's wife's place), *PaMaGumbo*, (a place of one of the *Gumbo*/Leg totem), *PaNdari* (the Shabeen), and *PaMusha* (at the homestead), which are open spaces, revelers gather over traditional meals and drinks "dialoguing" as well as discussing social, political, and economic problems that became characteristic of the Zimbabwean situation especially at the dawn of the twenty-first century ushered in by the radical land reform program.

In social conversations cited above, identification of each other and general "totemic" use to reflect individual ties is common, traditional and brings about a sense of belonging. Hence, *PaMaGumbo* (a place of one of the *Gumbo*/Leg totem) is no accident as it is purely a place name heavily enshrined with totemic connotations. There is no doubt that Africans believe extensively and express their totemic connections. No wonder the then President of Zimbabwe, Robert Mugabe, characterized Africans who opposed his rule as lacking "totems" in direct reference to people of European descent, who had migrated to Zimbabwe or expressed connections as well as ties to be Zimbabwean.

In terms of *PaMaGumbo,* therefore, those of the *Gumbo*/Leg totem, as well as those who may have been brought into contact with those of the Leg totem (by marriage or otherwise), find homage and accommodation in such spaces from which they can also call them home. One, therefore, reads a systematic as well as deliberate choice of designing, constructing, and naming of the places and spaces not only to reveal the village concept, but also to market the place using a great deal of the dictates of the African heritage. Hence, not only has the village been reconstructed and economized, but its values also such as totems, modes of relations, and communications are molded deliberately to bring economic value to owners of the spaces as well as the urban community at large.

KwaMaiguru/KwaGulez (my brother's wife's place) signifies hospitality. *Maiguru* (Shona for one's brother's wife) or in urban lingo "*Gulez*" symbolize hospitality especially in as far as food and drink are concerned. The places while serving modern dishes, bias is mostly on traditional African dishes or menus from which revelers through consumption of the food reflect easiness as well as feel at home in "*Maiguru's*" sanctuary. What is interesting here is also the fact that ownership of *KwaMaiguru/KwaGulez* and even *PaMaGumbo* are women, reflecting women's involvement as well as participation in economic establishments of these restaurants. *KwaMaiguru/KwaGulez* and even *PaMaGumbo* are feminine, hence communicating connections to women of Africa and their role in cultural communication and basic operations of African life in general. Hence, owners of these spaces deliberately name them ridding heavily on traditional relationships as bound up within and by the village concept.

To reveal aspects of traditional African architecture characterizing the new spaces and how these have been enriched, an appeal to the structures is worthwhile. Hence, the physical structures marking the spaces are important. The African village, traditionally, is characterized by the "hut"; in other words, the hut is made up of wooden poles and clay soils are used to patch up the walling poles. In terms of the roof, again wooden poles are used and they are grass thatched. The buildings are conically shaped, and a collection of such huts in a defined space creates a "village" as exemplified in **Figure 2a** and **b**.

In modern spaces, the same concept (as shown in **Figure 1**) is also being used where grass thatched buildings (but now in brick and cement walls) are now the common feature (see **Figure 2a** and **b**). The shape or designs of the buildings are conical, and the roofs are grass thatched as exemplified in **Figure 2a** and **b**.

In some lodges, for instance, "the village lodge" in Gweru (Midlands Province) offers overnight accommodation, and rooms are "huts" (although equipped with modern furniture and other utilities inside as shown in **Figure 2c** and **d**). One sees dotted huts within the physical spaces where the lodges have been built. The rooms are also named (rather than numbered) after either game or totems. This represents much of the traditional African village and its surroundings.

Second, the food served involves traditional meats, insects, grasses, and vegetables. Utensils used are wooden symbolizing much of the African village practices on food consumption.Meals come in to complete the ends of the architecture and designs The outdoor spaces of the lodges are designed as if to place patrons in a "bush" bringing them close to "nature" as shown by **Figure 2e**. One has a sense of the natural ambience and for Africans such set-ups trigger nostalgia.

Furthermore, entertainment provided in the same spaces as well as the general ambient reflects African traditions as well. Traditional music, dances, and performances are organized for revelers especially in lodges and those restaurants located at the outskirts of towns and cities. These three characteristics enrich the buildings

Figure 1.
Typical African village made up of conically shaped and grass thatched grass.

and their names chosen as the spaces named should have characteristics that support the totality of that which is found in a traditional village.

The reconstruction of the African village is, thus, observed from a number of fronts. First, there is a deliberate designing and construction of the new spaces. The "huts" are conical in shape and now being constructed using modern building materials in tandem with changes that have happened over the years. While traditionally poles were used, due to advances in technology, modern bricks, and cement are now being used for walling. However, the roofing is made up of treated poles and is grass thatched. The processes involved in upholding the architecture under discussion reveal costs implications at the same time bringing about some economic activities involved in the construction and maintenance of such spaces. To exemplify this, we note that suppliers of food stuffs to be prepared such as insects, traditional grasses (and vegetables), meats, and so on have taken an economic drive.

During the construction phases themselves, the grass is bought, and treated tree poles are supplied and used with modern building materials such as cement and bricks to create an African ambient. Traditional dance groups as well as musicians of traditional music such as *mbira* all participate for a fee. While we read culture and heritage being displayed and consumed, the same culture and heritage has been economized through a reconstruction of the African village in such a manner that locals have found economic opportunities from selling their culture, this time not only to foreigners as tourists but also to their own kin. Visiting such spaces for local revelers is like a journey back to the village from which the pressures of the urban landscape are quenched. Colonialism witnessed a huge influx of Africans migrating to the cities for economic reasons [6]. Such migrations with time resulted in Africans having to move between the city and rural home quiet frequently. As it were,

> *...others resided in the cities for many years. But even those who spent their entire working life in the city intended eventually to return to their hometowns. This is reflected in the visits that many of the more permanent migrants made to the village, usually around Christmas.* ([6], p. 62)

Figure 2.
(a) Round huts built with modern materials, (b) grass thatched huts for a lodge, (c) a view from the inside modern hut, (d) how some of the lodges are advertised, and (e) the ambience and outdoor set-ups of the lodges.

But for Zimbabwe, while van den Bersselaar's [6] views are valid, even in historic Zimbabwe, the land reform program further redefined and created more opportunities for a greater percentage of the population to migrate to the city mostly for economic opportunities. In some cases, there is a total disconnection of some Zimbabweans from their rural homes especially under circumstances where none of their parents are surviving who in most cases are the elderly who may have retired to their rural homes. Hence, visiting newly created villages in the cities is thus in some cases symbolic journeys to their villages as well as their past. Further engaging the above, Saidi [30] observed that,

> *Given that Africans have for long battled with cultural erosion and identity crisis, African people exhibit and continue to reflect effects of spiritual tension with their culture and environment. Their connections to "mother earth" have over the years been loosened. However, with vernacular architecture and the recreation of the "new" landscape, it is hoped that the feeling of reconnection and returning to the source can be awakened as the concretization of the meanings of nature and aspects from their natural environment communicated through objects that trigger visualization, representation and symbolization of their heritage can foresee creation of a place where buildings act as an extension of the interior as well as cultural and spiritual space. (p. 13)*

This reflects a great deal of aspects of heritage conservation, which is being done architecturally. Heritage can be both tangible and intangible, and there is a thin line that divides the two. Tangible heritage refers to the material objects that are seen or those that can be touched. On the other hand, "intangible heritage" generally refers to that heritage, which is usually experienced. According to "intangible heritage, it is often regarded as traditional culture that reflects the identity of a particular nation or group."

From the above submission, the architecture of the new spaces discussed above has become the pivot for the tangible and intangible heritage more so that the traditional village concept has come to be conserved in a unique way and continues to be used in contemporary times. The conclusion that Hărmănescu and Enache [31] make in as far as vernacular knowledge in architecture is concerned come in handy. They say that,

> *The past offers growth recourses for the future through the knowledge gained. In this case, the future should recognize the past value offering identity of those who use and transmit knowledge. To ignore the past, the vernacular knowledge of a place means to waste resources knowledge of a place. (p. 415)*

This speaks to efforts far removed from the increasingly and deliberate strategies in the twenty-first century "to preserve and conserve natural areas to prevent their degradation and to halt the innumerable attacks to which the industrial model of economic development has subjected the planet" ([31], p. 29). While the world has been so grossed with conserving "natural areas" by enacting legislature for instance, there has not been interest in coming up with the strategies to conserve some important aspects, especially intangible heritage in Africa. For Zimbabwe, therefore, the vernacular architecture has come to play this conservation role of culture as the landscape heavily came under construction within terrains far removed from natural sites where international legislature may force the country to protect the landscape. It is also important to note that what is being conserved here are not the buildings per se but the traditional architectural designs, activities, relations, and cultural identities of spaces, thereby promoting consumption as well as "visualization, representation, and symbolization" ([30], p. 13) of Africa's habitant heritage.

Shetabi [32] points out that built heritage has an important role to play in them expressing tangible and intangible values. In other words, buildings, based on their architecture, names and objects therein, are visual links to the past, revealing how communities evolve socially, technologically, and culturally. The above discussed buildings and spaces are proof to this line of thinking in Zimbabwe. These efforts should be encouraged because if the world is experiencing "the largest wave of urban growth in history" ([33], p. 1) and that by 2050 and that by 2050, 66% of the world's population will reside in urban areas [33], it also means that the entire world is currently experiencing challenges in terms of balancing cultural heritage conservation and the requirements of urbanization. Zhang et al. ([33], p. 1) are thus right to point out that

> *In this context, sustainable development, which is a complex, multidimensional, and evolving concept that encompasses environmental, economic, social, developmental, and cultural components, is highly valued worldwide.*

It follows, therefore, that Zimbabwe has begun to witness architectural maneuvers in seeking to balance environmental, economic, social, developmental, and cultural components in which mere citizens are taking a leading role as custodians of their heritage by turning their private spaces into cultural centers that house vernacular architecture.

7. Conclusion

Broadly speaking, the designs, constructions, and naming of lodges and restaurants after the FTLRP signify African traditional sensibilities recreated in the urban terrain. It further means that not everything has been lost in as far as cultural values are concerned among Zimbabweans. The "new" lodges and restaurants represent the African sensibilities and subsequently identity in a contemporary global village. In the same breath, we may, however, be forced to make an aside that if places, whose primary function is for business or money making, are constructed and named using traditional sensibilities, how far are the cultural values given center stage?

There is no doubt that constructions and naming of the places are also a marketing strategy where owners draw from their culture, and there is no sure way of ascertaining whether this is a purely economic drive, cultural drive, or both. Whether owners are conscious to cultural values or not, in what has been characterized by the likes of Ngugi [2] as Africa's problem being led by the culturally eroded and neo-colonial elites, there is evidence of attempts to recreate the traditional African family based on the village concept. From the nature of buildings and their names, it cannot be divorced that locals share and communicate the traditional past through physical objects, space, and the various aspects therein such as artifacts, foods, music, and dance.

Perhaps, this is so because there has been a realization that indigenous identity is important for the survival of any people. Hence, the repackaging of new spaces is not only for cultural survival, but also for economic redefinition supported by heritage through vehicles such as vernacular architecture and names of these public spaces like lodges and restaurants. These are efforts in Zimbabwe to "Africanize" the urban terrain. Therefore, urban buildings and their names as well as studies of the same indeed can help greatly in packaging indigenous knowledge systems for the survival of African heritage. We note that the urban landscape/spaces can actually store and communicate the heritage as well as history of the African people if systematic packaging and support is rendered to initiatives such as discussed above. We should be on the lookout, however, that at the center of it, all should be value exerted on culture and heritage since tagging monetary figures on traditional values may see traditions being consumed without pedigree desires for conservation, and this may be detrimental to that which has been left of the African heritage.

Author details

Umali Saidi
Communication Skills Centre, Midlands State University, Zimbabwe

*Address all correspondence to: saeedu2010@gmail.com

References

[1] Yankuzo IK. Impact of globalization on the traditional African cultures. International Letters of Social and Humanistic Sciences. 2014;**4**:1-8

[2] Ngugi WT. Decolonising the Mind: The Politics of Language in African Literature. Harare: Zimbabwe Publishing House; 1986

[3] Saidi U. Cracking the nonverbal code: Analysing cultural artefacts of the BaTonga in Zimbabwe [unpublished DPhil. thesis]. Gweru: Midlands State University; 2016

[4] Mungoshi C. Waiting for the Rain. London: Heinemann; 1975

[5] Soyinka W. The Lion and The Jewel. London: Oxford University Press; 1963

[6] van den Bersselaar D. Imagining home: Migration and the Igbo Village in Colonial Nigeria. The Journal of African History. 2005;**46**(1):51-73. DOI: 10.1017/S0021853704000015

[7] Hammer A, Raftopoulos B, Jensen S. Zimbabwe's Unfinished Business: Rethinking Land, State and Nation. Harare: Weaver Press; 2003

[8] Vambe MT. The Hidden Dimensions of Operation Murambatsvina in Zimbabwe. Harare: Weaver Press; 2008

[9] Tibaijuka AK. Report on UNDP Mission on Zimbabwe. New York: UNDP; 2005

[10] Fontein J. Anticipating the Tsunami: Rumours, planning and the arbitrary state in Zimbabwe. Africa. 2009;**79**(3):368-398. DOI: 10.3366/E0001972009000862

[11] Potts D. Making a livelihood in (and beyond) the African city: The experience of Zimbabwe. The Journal of the International African Institute. 2011;**81**(4):588-605

[12] Sadomba WZ. A decade of Zimbabwe's land revolution: The politics of the war veteran vanguard. In: Moyo S, Chambati W, editors. Land and Agrarian Reform in Zimbabwe: Beyond White-Settler Capitalism. Dakar: CODESRIA; 2013

[13] Moyo S, Chambati W. Land and Agrarian Reform in Zimbabwe: Beyond White-Settler Capitalism. Dakar: CODESRIA; 2013

[14] Mutodi P. Zimbabwe's Fast Track Land Reform. London: Zed Books; 2012

[15] Ikegami Y. The Empire of Signs: Semiotic Essays on Japanese Culture. Tokyo: John Benjamins Publishing; 1991

[16] De Saussure F. Course in General Linguistics. (Trans. Baskin W.). New York: McCraw-Hill; 1966

[17] Peirce CS. Collected Papers of Charles Sanders Peirce. Vol. 2. In: Charles Hartshorne C, Weiss P, editors. Elements of Logic. Cambridge: Harvard University Press; 1960

[18] Sebeok TA. Signs: An Introduction to Semiotics. 2nd ed. Toronto: University of Toronto Press; 2001

[19] Johansen JD, Larsen SE. Signs in Use: An Introduction to Semiotics. (Trans. Gorlee DL, Irons J.). London: Routledge; 2014

[20] Braga MLS, Mundy KMB. Semiotics: Science or method? Dispositio. 1981;**6**(17):109-117

[21] Champagne AR. Semiotics of poetry by Michael Riffaterre. The Modern Language Journal. 1979;**63**(4):229-230

[22] Hale C. Umberto eco takes semiotics to the masses. A Review of General Semantics. 2011;**68**(3):255-263

[23] Eco U. Postscript to the Name of the Rose. (Trans. by Weaver W.). New York: Harcourt Brace Jovanovich; 1983

[24] Eco U. The Island of the day before. (Trans. by Weaver W.). New York: Harcourt Brace Jovanovich; 1995

[25] Eco U. The Mysterious Flame of Queen Loana. (Trans. by Brock G.). Orlando, Harcourt; 2005

[26] Cunningham DJ. Beyond educational psychology: Steps toward an educational semiotic. Educational Psychology Review. 1992;**4**:165-194

[27] Driscoll MP, Lebow D. Making it happen: Possibilities and pitfalls of Cunningham's educational semiotic. Educational Psychology Review. 1992;**4**(2):211-221

[28] Dunbar-Hall P. Semiotics as a method for the study of popular music. International Review of the Aesthetics and Sociology of Music. 1991;**22**(2):127-132

[29] Appadurai A. Modernity at Large: Cultural Dimensions of Globalization. Minneapolis, London; 1996

[30] Saidi U. Heritage, semiotics and innovations: Architectural space, object-designs, meanings and implications in sustainable development. Social Semiotics. 2018. DOI: 10.1080/10350330.2018.1443584

[31] Hărmănescu M, Enache C. Vernacular and technology. In between. Procedia Environmental Sciences. 2016;**32**:412-419

[32] Shetabi L. Heritage conservation and environmental sustainability: Revisiting the evaluation criteria for built heritage. In: Paper Presented at the Australia ICOMOS Conference; 5-8 November 2015; Adelide, Australia. 2015

[33] Zhang Y, Zhang H, Sun Z. Effects of urban growth on architectural heritage: The case of Buddhist monasteries in the Qinghai-Tibet plateau. Sustainability. 2018;**10**:1593, 2-1518. DOI: 10.3390/su10051593

Section 4

Preservation and Restoration of Historic Buildings

Chapter 6

Assessment of Selected English Colonial Heritage Features for Sustainable Conservation in Kaduna Metropolis, Nigeria

Marcus Balah Ryal-Net, Bogda Prucnal-Ogunsote and Erekpitan Omoikhefe Ola-Adisa

Abstract

Heritage conservation is an avenue for human sociocultural value sustenance. This book chapter establishes the significant heritage values, determines the relative influential index of deterioration as well as assesses the suitable sustainable heritage conservation options within colonial town of Kaduna. Questionnaires were processed based on purposive sampling and administering 80 of them to building industry and allied heritage professionals, having 50 questionnaires that is about 62% dully filled and returned for analysis. The research instrument was validated using face and content analysis, while the Cronbach's alpha reliability test value for internal consistency of questionnaire instrument was established as being greater than 0.7. The study findings affirmed that historic factors were most significant in value of heritage, while government neglect is the most significant cause of deterioration. Correspondingly, heritage value criteria were considered as more significant than heritage deterioration agents in ascertaining the preferred sustainable conservation options. Conclusively, heritage conservation can be achievable by aptly establishing its significant value and undertaking timely deterioration assessment. This study recommended fashioning of conservation policy model by experts that will guide government agencies on conservation programs for sustainable heritages. The model will encourage local community sociocultural and economic partnership for tourism development.

Keywords: colonial, deterioration, heritage, sustainable conservation, value, Kaduna metropolis and Nigeria

1. Introduction

In most parts of the world, colonialism left behind footprints of legacies and heritages with amazing lessons and inspirations to learn and draw from. Although it can be argued that Africa had always had her indigenous knowledge that are mostly anchored on what is considered as 'intangible heritages' [1], colonial heritage features have however had great influence on the social, economic, political and cultural development of most African nations. Ever since then most scholars have tended to model African nations' sociocultural landscapes including the explanation for their causes along the heritages left behind by their colonial masters.

Britain, for example, has a number of bodies and commissions that ensure the preservation of heritage resources and sites. As part of environmental assessment in Britain, developers are mandated to investigate whether or not their development will affect any heritage site, ancient monument or listed buildings [2]. Analysts have often than not based their arguments on the logic that the landscape of a place gives an insight of what has happened there over a given time frame. According to Peter Benton, a principal preservation planner and a member of the American Institute of Architects, he argued that by looking at the geology and landscape of communities and other materials around, you can piece together the history of a place [3]. It is these perceptions that have caused governments across the world to initiate policies and programmes aim at preserving heritage resources and sites.

The heritage of a place, be it material or immaterial and whether indigenous or foreign, is highly valued because it provides invaluable insights into the changing dynamics [4]. In Britain, these bodies and commissions include the English Heritage, National Trust, Historic Buildings and Monuments Commission and English Nature. In many cities of the countries that were colonised, a number of structures were built for administrative and economic convenience by the colonial authorities and are labelled as the memorials of their past [5].

For Nigeria, the heritage landscape is replete with several features both (temporal and in-temporal cum spiritual). In this case the major tangible structures as stated by [6] include Presbyterian Church, Calabar (1846); First storey building, Badagry, Lagos (1845); Egedege N'okaro residence, Benin (1906); and Gobirau minaret in Katsina (Habe dynasty, 250 years ago); others are Tafawa Balewa Square and the National Theatre completed in 1976 for the Festival of Arts and Culture (FESTAC) of 1977. The old secretariat building of Lagos is considered the signature tune for the English colonial/revivalist architectural heritage conservation of the late nineteen and twentieth century and currently one of the listed building in Lagos state [7] (**Table 1**).

There are several other unique heritage sites, buildings and monuments that abound in Nigeria worthy of careful study. Buildings, monuments and sites features are a major part of our national heritage; therefore, could serve as important tourist attractions that reaffirm the need for this study in Kaduna metropolis.

Said et al. [8] argued that post independent Africa seems to have their colonial features ignored just as the colonialists did to those of Africa's past. For heritage documentation, some activities are critical: identifying heritage values, describing them and incorporating the often conflicting principal stakeholders [9, 10]. The identified heritage experts include specialists in social and economic values, tourism experts, environmentalists, historians, planners and developers [11, 12]. Kaduna being the main colonial town of the Northern Nigerian Protectorate had several buildings and monuments that left their mark on the townscape [13]. The colonial heritage features left behind in Kaduna needed to be conserved as has been done in most developed countries of the world since doing so is not only important because they provide a sense of identity in a fast-changing world for future generations but also represent the past history of a nation. Between 1999 and 2006, a number of English historic buildings were lost to the several infill activities that destroyed historic sites and buildings together with their value and the information they contained in Kaduna Government Residential Areas [14]. Although not all of these could have been conserved, most of them were products of high levels of sophistications, knowledge, tradition and history that in many instances took hundreds or thousands of years to develop. The destruction of these properties in most cases came with irreversible losses of the critical heritage identity (design, specifications and construction techniques) contained in them; it would be imperative to designate some of them as being historic and thereby could enable the enactment

S/nos.	Heritage feature	Pictorial schemes	Brief background	Remarks
	Gobirau Minaret, Katsina (1348–1408 as earliest date)		Gobirau Minaret is a historical monument and landmark in Katsina. It was reported to have been built during the reign of Sarki Muhammadu Korau (1348–1408 AD) who was the first Muslim king during the Habe dynasty. Other account argued that it was built during later years, which is contestable considering the enormous support for the later date. The minaret is about 50 feet in height and was originally built to serve for calling Muslim faithful to prayers. However due to its height, it served the strategic purpose of sighting approaching enemies forces	There is this continuous debate about whether this is the first storey building or the Badagry building. While the Gobirau was mainly as a minaret and watch tower, it was not for habitation unlike the Badagry building which was occupied
	Presbyterian Church (1846)		On 10th April, 1846, the pioneer missionary team arrived, Calabar, and set up the first Presbyterian Church under the leadership of Rev. Hope Masterson Waddel. It was then known as the Presbytery of Biafra. The Presbytery of Biafra evolved in 1921 into Presbyterian Synod of Biafra. By 1945, it was handed over to Nigerians, and the completed mission church integration was achieved on 16th June 1960. Thus becoming the Presbyterian Church of Nigeria	It is worthy to note that this mission station came into being at the request of the local elders after some long period of interaction with the sea-faring European traders
	First storey building, Badagry Lagos (1845)		The first storey building in Nigeria had its foundation laid in 1842 and completed in 1845 located at Badagry, Lagos. The building was said to be built by a Church Mission Society (CMS) reverend gentleman named Henry Townsend. The challenge faced by the structure is the urgent need for rehabilitation towards conserving the heritage feature for future generation. Most visitors come to visit the building during important festival around August of each year	The first storey building in Nigeria has remained the oldest known inhabited building in Nigeria, while the Gobirau minerate is a monument of the past Habe dynasty in Katsina

S/nos.	Heritage feature	Pictorial schemes	Brief background	Remarks
	Egedege N'Okaro, Benin City (1903–1905)		The first storey building in Benin city. Its construction was started in 1903 and completed in 1905 by Chief Iyamu, the Ine of Benin. It was a great fit and quite awesome structure compared to its surrounding mud hut landscape	The Egedege N'Okaro is a most visit for any tourist to the Mid-Western Nigeria; particularly for visitors who want to explore the great Benin city wall and moat remains; exploring this residence of over a century is worthwhile
	Old Secretariat building, Lagos (1906)		The Old Secretariat building in Lagos Island was designed and constructed by the Public Works Department (PWD) during the colonial era at the declaration of Lagos as the capital city of the Southern Nigeria Protectorate in 1906. It served as the office for the colonial secretary and accommodated several other ministries. The secretariat is single bank with large corridors, has expansive windows and is structurally made up of cast iron, timber post and brick walls. The building was declared a national monument on 6th August, 1982, by the National Commission for Museum and Monuments	The nature of design and material used for construction provides cover against driving rain along the corridors. The room temperature is also controlled by the wide opening and thick walls further shielded by overhanging roof
	Tafawa Balewa Square, Lagos		Tafawa Balewa Square was initially a race course harbouring memorials for World Wars I and II as well as the Nigeria Civil war. The square occupies about 14.5 hectares (35.8 acres) and was built in 1972 on 45/57 Massey Bamgboshe Street, Lagos. The major unique feature of the square is the four white horses and seven eagles hovering over the entrance façade columns and rooftop, respectively. The square was the venue for the Nigerian independence celebration in 1960	At full capacity, the square can accommodate up to 50,000 spectators at once. It is a major social-gathering place for festivals and ceremonies of note within the community

S/nos.	Heritage feature	Pictorial schemes	Brief background	Remarks
	National Theatre, Iganmu Lagos (1976)		The National Arts Theatre Iganmu, Surulere, Lagos, has remained the dominant centre for Nigerian performing arts since its construction in 1976. It was built mainly for the hosting of Africa's second festival of arts that was held in Nigeria in 1977. Is a major recreational and tourist destination in once in Lagos, Nigeria	The design concept of the theatre was based on the Bulgarian cultural and sport palace

Source: Variously adapted from [6, 7].

Table 1.
Selected heritage buildings across Nigeria.

of appropriate preservation ordinances to protect these significant buildings and monuments before a worst fate befalls them.

This study examined experts' judgement of perceptual values of key colonial structures in Kaduna metropolis which include offices, residences and symbolic monuments. The study takes into consideration key approaches to sustainable heritage conservation which includes preservation, repairs, maintenance, restoration, treatment, dismantling and relocation as well as reconstruction in consultation with users and local communities [15, 16]. Deterioration agents were associated with government neglect; tear and wear; poor level of awareness; vandalism; chemical, thermal, biological and condensational activities; and mechanical and electromagnetic solar effects [9, 12]. However, since value is an underlying rationale for heritage conservation, value perception was restricted to experts in arts history, archaeologist, architects and others who have been identified as critical fresh stakeholders or constituents [17].

Apart from the empirical need to conserve heritage buildings and monuments in Kaduna, studies of colonial features have not had their fair share in the post independent African heritage studies. The need to increase the city's building and monument profile in the country's stock of listed buildings has received attention in recent times. Zubairu et al. [18] opined that as historic as Kaduna metropolis is in the community of historic cities in Nigeria and Africa as a whole, only Lord Frederick Lugard Bridge in Kaduna was a listed structure for Nigeria. This leaves much to be desired considering the vast heritage features that abound in Kaduna as the main colonial town of Northern Nigeria [13]. All these realities call for concerted efforts to improve the image of heritage structures in Nigeria and Kaduna in particular including the necessary tools to make it work from the perspective of Nigerian realities.

The lack of listed building and monument heritage particularly colonial features in Kaduna metropolis is of great concern. Therefore, value perception assessment of significant historic buildings and monuments is therefore a rational step towards sustainable conservation of these properties. The study similarly examined and tackled the fundamental issue of methodological challenges associated with value assessment of historic buildings and monuments as well as the need to contextualise the heritage conservation and tourism need of Kaduna metropolis as a colonial city.

1.1 Objectives of the study

The specific objectives of the study are as follows:

a. To establish the perceived heritage value significance for sustainable building and monuments conservation in Kaduna metropolis

b. To determine the perceived effect of deterioration caused by various agents and environment on the heritage buildings and monuments within Kaduna metropolis

c. To assess the most suitable sustainable conservation options that enriched colonial heritage buildings and monuments within Kaduna metropolis

2. Review of relevant literature

A further review of relevant variables for this study is undertaken to aid in contextualising the specific variables of the study.

2.1 Conservation of heritage

Heritage and conservation are interrelated in a way, as heritage is considered an expression of the sociocultural value product of a people, its conservation ensures the bequeathal to posterity. The research team [19] stated that conservation serves as a medium for architectural and historical heritage revitalisation. Meanwhile, to revitalise the heritages, it is significant to identify and assess the various relationships involved. In this study the relationships are with respect to heritage value, deterioration and their prospect for sustenance. Similarly, [20] contended that conserving urban heritage is growing with better funding; however, he argued that the partial perception of heritage conservation equally has varied scopes. On the other hand, [15] contended that conservation is categorised based on the various physical buildings and spatial relationship. [15, 16] did affirmed that various approaches for sustainable heritage conservation are applicable and include preservation, repairs, maintenance, restoration, treatment, dismantling and relocation as well as reconstruction in consultation with users and local communities. Likewise, [21] maintained that cultural heritage features are made up of buildings, monuments, gardens, cemeteries, landscape and archaeological sites.

2.2 Conservation and sustainability issues

Conservation of heritage building or site is a process of catering for a place so that it can retain its significant historical, architectural, aesthetic and cultural value. Similarly, Gro Harlem Brundtland Commission Report 'our Common Future' in 1987 defined sustainable development that is succinctly paraphrased in [22] as '... meeting the needs of present generation without jeopardizing the ability of future generation to meet their needs'; [23] further reinforced the argument that conservation relates directly to sustainability, while [24] also averred that sustainability is centred around economic, social and environmental parameters.

Studies by [25, 12] as well as [26] variously argued that heritage conservation should incorporate public interest, sociocultural values, as well as environmental parameters. Often the terms sustainable development, conservation and sustainability emphasise connection amongst social equity, economic production and environmental quality, which should be adaptable to heritage setting.

Kaduna as a main Gbagyi settlement was transformed into a major town for all Northerners that housed its colonial-built treasures. Kaduna metropolis was initially settled by mainly colonialist, their dependents, workers from within West Africa and Southern Nigeria Protectorate [5]. Towards entrenching a functional colonial system, structures were built that include private residences, offices, services and symbolic structures as bridges and entrance gates which form the fulcrum of this particular study.

2.3 Heritage buildings and monuments

The International Council on Monuments and Sites (ICOMOS) Venice Charter of 1964 adopted during the 2nd International Congress of Architects and Technicians of Historic Monuments made definite decisions on heritage. During the congress it was advocated that heritage should maintain buildings and monuments to ensure social usefulness, within original scale and historical setting [27, 28]. Furthermore, ICOMOS Charter of 2003, at Victoria Falls, Zimbabwe, argued that structures of architectural heritage lay confronts for restorative finding in terms of nature and history suitable for the cultural context. During the 14th general assembly at Zimbabwe in 2003, they also establish the principles that cover general criteria, research, diagnosis, remedial measures and controls. Specifically, the principles' general criteria cover conservation, reinforcement and restoration; others are value and authenticity, integrity, functional changes, peculiarity of heritage structure and the underlining concept of no action to be taken until benefits can be ascertained proactively. EngleHardt [25] argued that built heritage provides the physical space for the nonphysical expressions of culture and constitutes one of the most varied, complex and eloquent manifestations of tangible cultural heritage. Here then is the essence of this particular study towards conserving Kaduna colonial heritage features with specific emphasis on its buildings and monuments.

2.4 Heritage values and deterioration

For any heritage to be conserved for control or minimised decay or deterioration, there is the need to establish its value to the community or society at large. Morgan et al. [29] opined that value is in general the same as meaning or in valuation, meaning confirmed and enhanced with emphasis. Value is here considered as the result of brain conversion of direct experience (immediate perception) into meaning and actionable truth (a secondary judgement or conception). The Allen Consulting Group [30] contended that historic heritage features in Australia are priceless in terms of their value. It is here believed that there is value creation for architectural designs and literature through heritage conservation. This was further elaborated by [31] as they considered heritage value as mainly based on use state rather than the exchange rate. Heritage value in this context has to do with historical, architectural, political, sociocultural, technological and economic value of the heritage features within Kaduna metropolis.

Building deterioration on the other hand has to do with its material component or finishes that do not meet its initially acceptable performance criterion. Mydin et al. [12] stated that often buildings suffer from defects that range from cracks, water penetration or human activity-based deterioration. Factors considered as responsible for deterioration of heritage buildings are broadly classified as natural (environmental) and man-made (social) [9, 12, 32]. It is these key issues that underlay the concern of this paper and have been articulated for enhanced perception and applicability.

3. Research methodology

The study heritage features are mainly along Ali Akilu way through Ahmadu Bello way to Junction road as well as Independence way linking back to Junction road. The choice of these areas/routes was due to their historical place as the core segment of the town during colonial era. These routes also harbour virtually all the colonial administrative features of the 1900 to 1960 period in Kaduna metropolis. Stratified purposive sampling of building heritage and monuments within Kaduna metropolis was made. With the aid of convenience sampling method, a sample design of selected heritage and allied professionals was made [33]. A total of eighty (80) questionnaires were administered to various respondents in the built environment and experts in heritage and conservation within Kaduna metropolis. Fifty (50) of the questionnaires were duly filled and returned with the help of four research assistants, while three (3) were returned but not duly filled thus were considered invalid. This will aid in ascertaining the respondent perception in respect to value, deterioration and suitable sustainable conservation.

Feedback on the questionnaire was provided on its appropriateness as well as information presented (face and content validity test) to ascertain whether the main issues were effectively communicated [34]. The questionnaire instrument employed used internal consistency reliability to assess the results across the different items [35]. The instrument was shown to have value range of 0.788 for deterioration, 0.876 for sustainable conservation options and 0.903 for heritage value using Cronbach's alpha reliability test. Therefore, since the Cronbach's alpha values are all greater than 0.7, it shows that the instrument is reliable for obtaining information from respondents [24, 36].

For this study reflective influence index (RII) tool was used to analyse objectives as in Somiah, Osei-Poku and Aidoo (2015). Below is the formula for the various analyses as utilised for this particular data set collated from respondents:

$$RII = \frac{1n_1 + 2n_2 + 3n_3 + 4n_4 + 5n_5}{5(n_5 + n_4 + n_3 + n_2 + n)} \times \frac{100}{1}$$

where RII (%) is the relative importance index of each factor, while n_1, n_2, n_3, n_4 and n_5 are the numbers of observed frequency. According to the research instrument, '1' represents highly insignificant value (HIV); '2' represents insignificant value (IV); '3' represents moderately significant value (MSV); '4' represents significant value (SV); and '5' represents highly significant value (HSV).

4. Results and analysis

The colonial administration core buildings identified and used in this study are Hassan Katsina (State) House, Lugard Hall building, Lord Lugard Residence/Water Tower, Secretariat Complex and Magajin Gari. The monuments on the other hand are the Northern Entrance City Obelisk, Kaduna Railway Junction Steel Bridge and the Lord Lugard Bridge.

4.1 Selected heritage buildings and monuments in Kaduna metropolis

Below is the description of these heritage buildings and monuments stating their basic features with pictorial illustrations.

Hassan Katsina State House Complex, along Kawo way, was an administrative office building used since the colonial period and currently serves as the office of His Excellency the Deputy Governor of Kaduna state. The feature within the

Figure 1.
Hassan Katsina State House with banquet hall at the background.

complex as you enter the main gate is the banquet hall that serves as the state for major occasions (**Figure 1**). Its design is an integration of the traditional Hausa Architecture and utilisation of colonial-styled materials and skill craftsmanship for its design and construction.

Kaduna Commemorative Entrance Obelisk is mainly made up of well-carved stoneworks at the Northern entrance of the city just by the entrance of the old Nigerian Defence Academy (NDA). It was erected to commemorate the visit of Her Royal Majesty Queen Elizabeth and His Highness the Duke of Edinburgh to Kaduna in February, 1956 (**Figure 2**). Recently it was demolished and relocated to allow for the expansion of the major road artery to the town from the northern axis from 4 to 6 lanes. The initial concept was maintained, and the old stones were also utilised with additional stoneworks crafted in similar fashion as those of the initial commemorative entrance monument.

Lord Lugard Hall is one of the major symbols of the great history of Kaduna town from colonial times to post independence. It served as the house of chiefs and later translated to the present Kaduna State legislators House of Assembly (**Figure 3a** and **b**). It is strategically located just off Ali Akilu road along the old Coronation road presently Independence way that lings to the State Secretariat Complex.

Lord Lugard Residence/Water Tower is located off Coronation Avenue present day Independence way just adjacent the Lugard Hall; it is currently vandalised and laid bare on the pretence of government rehabilitating the historic residence. The house is a duplex situated on an expand of land that has been subject to several litigation towards being excised (**Figure 4a** and **b**). The last administration in the state gave out the contract for the rehabilitation of the Lugard residence; unfortunately it became a pretext for vandalised very antique materials available in the building. This was more serious with respect to doors and windows including ceilings and the roof structure/lining. Recently, the current government has commissioned another set of consultants to produce a proposal towards restoration and adaptation for museum with prospect as a major tourist feature for the town. Just beside the residence is the Water Tower that is used to serve the Lugard complex and often used as a watch tower during the early days of Kaduna city development shown in **Figure 4c**.

Figure 2.
Kaduna Entrance Obelisk at old NDA Junction around Federal Court of Appeal.

Figure 3.
(a) Lugard Hall currently serving as Kaduna State legislative House of Assembly and (b) Lugard Hall entrance gate.

State Secretariat Complex along the Independence way is a collection of very interesting and unique colonial heritage buildings that housed the Civil Service Ministries then and now. The ministries currently being housed include Education, Economic Planning and Finance (**Figure 5a** and **b**). **Figure 6** is showing the ministry of Justice and Ministry of Water Resources. Across the road directly opposite these ministerial complexes are two other ministries. They are the Ministry of Health and Women Affairs in **Figure 7a** and **b**. Most of the buildings have undergone some form of renovation, particularly the replacement of roofing structure and linings. The main building structure is still intact and wonderfully preserved in its initially built form.

The Magajin Gari Complex is made up of the Local Government Secretariat and the Sharia courts (**Figure 8a** and **b**). It was formally the centre for the Northern Nigeria Native Authority administration. The architecture is distinctively made up

Figure 4.
(a and b) Lord Frederick Lugard residence as you approach and the roof trusses showing well-seasoned timber bolted and secured for about a century. (c) Lord Frederick Lugard residence Water Tower often used as a watch tower during early days of Kaduna township development.

Figure 5.
(a) Ministry of Education at the background and the Ministry of Economic Planning and Finance at the foreground and (b) the stairs being renovated at the Education section.

of traditional Hausa form, domes, arches, decorative motifs and varied pinnacles (Zankwaye). These pinnacles were an extension of the parapet walls in a traditional building and are basically symbolic manifestation of aesthetic elegance in traditional architecture. These became so because they have lost their initial function as stake around which master builders dangle rope ladder for climbing roof or plastering of building facades.

Lugard Bridge is a very important steel bridge historical monument that was first constructed in 1904 at Zungeru but brought into Kaduna when the colonial master relocated the Northern protectorate headquarters to Kaduna (**Figure 9a** and **b**). It is

(a) (b)

Figure 6.
(a) Corridor along the Ministry of Water Resources and (b) Ministries of Justice at the background and Water Resources at the foreground in need of renovation.

(a) (b)

Figure 7.
(a) Kaduna state Ministry of Health and Human Resource building and (b) Kaduna State Ministry of Women Affairs building at the foreground.

(a) (b)

Figure 8.
(a) The Kaduna North Local Government Area building and (b) Shariah Court of Appeal as part of Magajin Gari Native Authority Administrative building.

presently sited at the old Kaduna Garden now Hassan Katsina Park. Was re-erected in 1954 and has been there since then along the swimming pool road beside the old shooting range.

Kaduna Railway Junction Steel Bridge, over river Kaduna, is linking the two parts of Kaduna metropolis (North and South). It is quite a unique view of excellent steel works on concrete columns running parallel to the main arterial road linking

(a) (b)

Figure 9.
(a) Lord Lugard Bridge at Gamji Gate showing the steel frame across one of the streams to river Kaduna and (b) bridge showing the timber planks serving for foot passage.

(a) (b)

Figure 10.
(a) Railway line bridge across river Kaduna and (b) intricate steel works carrying the rail line across river Kaduna.

the two parts of the city made up of concrete bridge as you journey daily from one end of town to the other (**Figure 10a** and **b**).

4.2 Analysis of heritage value significance in Kaduna Metropolis

Table 2 displays RII values of Kaduna heritage building and monument value significance; it was suggested that heritage buildings were ranked the first with significant influence on the heritage value, followed by heritage monuments. Similarly, historical factors were ranked the highest variable under heritage building and monument categorisation, being critical variable that influence the heritage value of heritage building and monuments and economic factors as being the least significantly valued by respondents.

4.3 Analysis of agents that cause building and monument deterioration

Table 3 revealed that government neglect being ranked the highest is considered the most significant agent that causes building and monument deterioration, followed by tear and wear of building and monument deterioration; then lack of public awareness; vandalism; chemical, thermal, biological and condensational activities; and mechanical and electromagnetic solar effects. The least agents that significantly influence the deterioration of buildings and monuments are considered as electromagnetic solar action.

	HIV	IV	MSV	SV	HSV	RII (%)	Rank
Heritage building						71.4	First
Historical	3	8	1	9	29	81.2	
Architectural	1	6	10	16	16	76.3	
Political	4	7	5	19	15	73.6	
Sociocultural	1	8	13	21	7	70.0	
Technological	3	13	10	17	7	64.8	
Economic	4	11	17	10	8	62.8	
Heritage monuments						68.7	Second
Historical	4	4	7	12	23	78.4	
Architectural	1	10	8	18	13	72.8	
Political	1	9	18	17	5	66.4	
Sociocultural	2	9	18	14	7	66.0	
Technological	5	6	14	19	4	64.6	
Economic	6	6	18	13	7	63.6	

Note: highly insignificant value, HIV (1); insignificant value, IV (2); moderately significant value, MSV (3); significant value, SV (4); and highly significant value, HSV (5).

Table 2.
RII of Kaduna heritage building and monument value significance.

Agents that cause deterioration	HI	IV	MS	S	HS	RII (%)
Government neglect	3	2	3	15	27	84.4
Tear and wear of usage	2	2	7	19	19	80.8
Lack of public awareness	1	4	6	21	17	80.0
Vandalism	3	9	6	14	17	73.5
Chemical	1	8	18	6	15	70.8
Thermal	0	8	18	9	12	70.6
Biological	3	8	18	9	12	67.6
Condensation activities	2	13	11	16	8	66.0
Mechanical	2	11	15	10	9	65.5
Electromagnetic solar	3	16	12	9	9	62.0

Note: highly insignificant, HI (1); insignificant, I (2); moderately significant, MS (3); significant, S (4); and highly significant, HS (5).

Table 3.
RII of agents that causes building and monument deterioration.

4.4 Options for sustainable conservation that engender heritage buildings and monuments

Table 4 depicts the RII values of options for conservation that engender heritage buildings and monuments in Kaduna state. It is determined that heritage building ranked first in significantly influencing the conservation works, followed by heritage monuments. Renovation was ranked the highest variable under heritage building as critical variable that influence the conservation works of heritage.

Heritage and conservation options	HI	I	MS	S	HS	RII (%)	Rank
Heritage building						75.4	First
Maintenance and treatment (renovation)	2	4	1	15	27	84.9	
Repairs of damaged sections	1	5	6	17	21	80.8	
Preservation (maintain fabric)	5	4	4	12	25	79.2	
Restoration (return to earlier known state)	2	6	9	7	23	78.3	
Adaptation (change function)	2	9	16	12	6	64.9	
Reconstruction	6	9	12	12	8	63.0	
Heritage monuments						72.2	Second
Preservation (maintain fabric)	2	4	9	13	22	79.6	
Maintenance and treatment(renovation)	3	3	9	15	19	78.0	
Repairs of damaged sections	3	4	12	13	18	75.6	
Restoration (return to earlier known state)	3	10	9	7	21	73.2	
Reconstruction	7	9	11	9	11	63.4	
Adaptation (change function)	2	13	13	15	4	62.6	

Note: highly insignificant, HI (1); insignificant, I (2); moderately significant, MS (3); significant, S (4); and highly significant, HS (5).

Table 4.
Options for conservation works of heritage buildings and monuments.

Preservation (maintain fabric) is ranked highest variable for heritage monuments, being a critical factor that influenced the conservation works based on established value and deterioration effect on heritage monuments. Reconstruction and adaptation (change in function) for buildings and monuments, respectively, were considered the least options for heritage conservations of English colonial buildings in Kaduna metropolis.

5. Findings and discussions

The study findings and discussions can be surmised as:

1. Heritage buildings were considered more significant in value than monuments; this could be mainly due to utilitarian perception by respondents than valuation. In both buildings and monuments, the historical value is considered the most significant as these features are basically from the colonial past and being seen and considered as such, while the economic valuation was seen as least significant as most of these structures currently are not particularly being use for any financial benefit. In fact, even the railway infrastructures which used to be the economic nerve of the colonial era is no longer being seen as such. There will therefore be need for strategic exposure of younger generation to some of these colonial heritage features and their historical significance if these features will have a future (Scerri, Edwards and Foley, 2016).

2. The major causes of building and monument deterioration were established as being government neglect which could be based on the fact that all of the structures are currently serving various government agencies. Meanwhile,

considering the paucity of funds, conservation of heritage is not given appropriate consideration based on government scale of preference. On the other hand, the minimal deterioration caused by electromagnetic and solar effect on monuments could be attributed to the nature of most of the feature design and specifications. In the design there was extensive provision of veranda, corridors and recesses on most openings; as well as high-quality and durable materials were specified and used during construction. While the government is considered the most critical culprit in heritage deterioration, the other side of the challenge will be that most other stakeholders may stand aloof without bringing to bear their expertise as well as communal consideration in conserving the features.

3. Heritage buildings have higher relative influence index of 75.4, with monuments having 72.2 on sustainable conservation options. Similarly, maintenance and treatment (renovation) conservation option have an index of 84.9 for buildings and 79.6 for monuments in Kaduna metropolis.

4. The option for heritage building conservation has the highest reflective influence index (75.4) since most respondents are more conversant and interact with the buildings more frequently than they do with monuments (72.2). In fact, most monuments within the study area are either abandoned or have no active significance in the daily life of the city dwellers, therefore making them seem less attractive for conserving than the buildings that are currently being used as offices and other administrative accommodation. The building heritage maintenance and treatment (renovation) was most significant at 84.9 option for conservation as most of the buildings materials are dying and rarely in used. Currently some of them are being demolished and redeveloped using less responsive design and specification options, hence the preference of professional for this option. The Monuments most preferred option for conservation was considered as preservation (maintaining of the fabric) at 79.6 since such materials are mostly not available or being used in such construction currently. Thus the significance of preserving the heritage features for posterity is towards serving as major tourist site now and in the future. Reconstruction of buildings was considered the least option for conservation and adaptation (change of function) for monument as such action could easily undermine the significance of the heritage features and their place as historical legacies. The respondents' perception of the building value is core to any conservation work than the features' level of deterioration. This further affirms the place of value in heritage conservation as variously argued (Mydi, Ramli and Awang, 2012).

6. Conclusion and recommendations

In conclusion, the study was able to ascertain the place of heritage buildings and monuments for colonial Kaduna metropolis. This was done through integrating the key heritage value indicators and deterioration agents to determine most suitably significant heritage sustainable conservation options. By this, the study set in motion an evolving trend of value rating and deterioration assessment model. This model if adopted can enable forecasting suitable conservation options for enhanced tourism activities as well as points direction for prospective architecturally and historically significant developments. Here from further detail, studies could be undertaken to ascertain specific heritage building and monument value as well as their status of deterioration for the various stakeholders' appropriate action. More purposeful sustainable heritage buildings and monuments are attainable by careful conservation measures on each heritage structure depending on its requisite need.

Based on study finding, the following recommendations are hereby presented as follows:

1. Heritage experts are to evolve acceptable value significance factors for heritage buildings and monuments in Kaduna metropolis for colonial buildings.

2. Government agencies occupying heritage buildings should be mandated to produce and adhere to professionally developed maintenance manual for all buildings and monuments with the experts' inputs.

3. Historic relevance of each monument should be periodically highlighted to further enlighten people on their significance so that heritage monuments conservation can be brought to the fore.

4. Adopt a comprehensive conservation template for all heritage features as an integral metropolis tourism package for possible generation of financial resources that can be used for conservational activities.

Finally, this submission supposes that a more sustainable heritage conservation can be attained and maintained with periodic value reassessment as well as establishment of actual deterioration status of each feature for the benefit of posterity.

Author details

Marcus Balah Ryal-Net*, Bogda Prucnal-Ogunsote and Erekpitan Omoikhefe Ola-Adisa
Department of Architecture, Faculty of Environmental Sciences, University of Jos, Nigeria

*Address all correspondence to: ryalnetmarcus@gmail.com

References

[1] Munyaradzi M. Indigenous knowledge and public education in sub-Saharan Africa. African Spectrum. 2015;**50**(2):57-71

[2] CIRIA. Guide to the Management of Building Refurbishment: Report 133. London: CIRIA; 1994. Retrieved from: https://www.shu.rl.tallis.com/

[3] Michael B, Nancy F, Jason V. Becoming and Urban Planner: A Guide to Careers in Planning and Urban Design. New Jersey, USA: John Wiley & Sons Inc; 2010

[4] Bendix RF. Patronage and preservation: Heritage paradigms and their impact on supporting "Good Culture". In: Adell N, Bendix RF, Bortolotto C, Tauschek M, editors. Between Imagined Communities and Communities of Practice: Participation, Territory and the Making of Heritage. Vol. 8. Gottingen: Gottingen University Press; 2015. pp. 219-234

[5] Bununu YA, Ludin AN, Hosni N. City profile: Kaduna. Cities. 2015;**49**:53-65. Available from: https://www.researchgate.net/publication/280614102city_profile_Kaduna

[6] Wahlay H. Top 10 historical buildings in Nigeria. Nairaland Forum. 2015. Available from: www.nairaland.com/2142894/top-10-historical-buildings-Nigeria

[7] Opara C. The old secretariat: In need of revival. Ekotecture Quarterly. Online Magazine for NIA Lagos Chapter. 2017

[8] Said H. The history and current situation of cultural heritage care in sub-Saharan Africa. Asian and African Studies. 1999;**8**(1):91-100

[9] Mostafa SL. Historical Cairo: Causes of deterioration and suggestion for preservation [Internet]. nd. Available from: www.ICOMOS.org/publications/wash101 [Accessed: 20 May, 2017]

[10] Torre M, Mason R. Assessing the Values of Cultural Heritage. Los Angeles: Getty Conservation Institute; 2002

[11] DPIPWE. Assessing Historic Heritage Significance, Version 5. Tasmania, Australia: Department of Primary Industries, Parks, Water and Environment; 2011

[12] Mydin MAO, Ramli M, Awang H. Factors of deterioration in building and the principles of repair. ANUL XIX, NR.1. 2012. Available from: Anale-Ing-Uem.Ro/2012/39.Pdf

[13] Qurix WB, Suleiman E. Transformation and continuity of architecture in Kaduna city, Nigeria, from 1900-2000. Canadian Open Environment and Earth Science Journal. 2015;**1**:1-16. Available from: http://www.crpub.com/journals.php [Accessed: 03 August, 2017]

[14] Arinde SH. An Analysis of Patterns and Implications of Infill Development in GRAs of Kaduna. Zaria: Department of Urban and Regional Planning, Ahmadu Bello University; 2006

[15] Hadri K, Boussaa D. Architectural and urban conservation in the United Arab Emirate. Open House International. 2007;**32**(2):16-26

[16] Harun SN. Heritage building conservation in Malaysia: Experience and challenges. Procedia Engineering. 2011;**20**:41-53. During the Second International Building Control Conference

[17] Jokilehto J. A History of Architectural Conservation. Oxford: Butterworth-Heinemann; 2018

[18] Zubairu SN, Abdulrahman ME, Ayuba P, Adedayo OF. A study of listed buildings and monuments in Nigeria (1956-2006). Journal of Economics and Sustainable Development. 2012;**3**(7): 89-99. Available from: http://www.iiste.org [Accessed: 20 Apirl, 2017]

[19] Research Team. Conservation and revitalization of historic buildings. Liberal Studies Teaching Kit for Senior Secondary School Curriculum. Community Project Workshop, Faculty of Architecture, The University of Hong Kong

[20] Rojas E. The conservation and development of the urban heritage: A task for all social actors. City and Time. 2007;**3**(1):4. Available from: http://www.ct.ceci-br.org [Accessed: 23 May, 2017-05-23]

[21] CPWD. A Handbook on Conservation of Heritage Buildings. New Delhi, India: Publish by Directorate General Central Public Works Department; 2013

[22] World Commission on Environment and Development. Our Common Future. Oxford, UK: Oxford University Press; 1987

[23] Theis T, Tomkin J, editors. Sustainability: A Comprehensive Foundation. Houston, Texas: Rice University, Connexions; 2012

[24] Allu ELA. Climate change and buildings in Nigeria: A search for mitigation and adaptation framework for residential design guide [an unpublished thesis]. Leicester, United Kingdom: Partial Fulfilment of the Requirement for the Doctor of Philosophy in De Montfort University; 2014. Submitted

[25] EngleHardt RA. "First principles" in conservation: The UNESCO Asia-Pacific heritage award. APT Bulletin: Journal of Preservation Technology. 2010;**41**(2-3):57-65

[26] Fisher T. Ethics and architectural practice. In: Hayes RL, editor. The architect's hand Book of professional practice. 15th ed. New Jersey, USA: American Institute of Architects: John Wiley and Sons, Inc; 2015

[27] ICOMOS. International Charter for the conservation and restoration of monuments and sites (The Venice charter), Paris, France: ICOMOS. 1964; Available from http://www.icomos.org/charters.nara-e.pdf

[28] Zaryn A. "The first general assembly of ICOMOS 1965 Cracow Poland: Regulations, bylaws and National Committees" in thirty years of ICOMOS. Scientific Journal. 1995;**25**:3-6

[29] Morgan MH. Vitruvius: The Ten Books of Architecture. London: Harvard University Press; 1914

[30] The Allen Consulting Group. Valuing the priceless: The value of historic heritage in Australia. Research Report 2. Sydney: Heritage Chairs and Officials of Australia and Newzerland; 2005

[31] Edwards R, Holland J. What Is Qualitative Interviewing? London: Bloomsbury; 2013

[32] UNESCO. Preserving and Restoring Monuments and Historic Buildings. Paris: United Nation Educational, Scientific and Cultural Organization; 1972

[33] Suri R, Rao J. Impact of spiritual marketing on different segments of tourist and their evaluation of the site. Journal of Bussiness and Eeconomic Policy. 2014;**1**(1):26-34

[34] The Allen Consulting Group. Boat industry training market analysis and conceptual model. In: Report. Melbourne: Department of Education and Early Childhood Development; 2013

[35] Gusen JN. Information and Communication Technology. Department of Science and Technology, University of Jos; 2017

[36] Travakol M, Dennick R. Making sense of Cronbach's alpha. International Journal of Medical Education. 2011;**2**:53-55